Stoic Philosophy and the Control Problem of AI Technology

Values and Identities: Crossing Philosophical Borders

Series Editors: Paul Crowther is Professor of Philosophy at the National University of Ireland, Galway, Tsarina Doyle is Lecturer in Philosophy at the National University of Ireland, Galway

How do values define human identity and the different activities through which this identity finds expression? *Values and Identities: Crossing Philosophical Borders* publishes research-led monographs and edited collections that face this problem head-on. Titles in this series investigate specific forms of value and, in particular, how they interact across societal contexts to form more complex identities.

Titles in the Series
Virtue as Identity: Emotions and the Moral Personality, Aleksandar Fatić
Human Value, Environmental Ethics and Sustainability: The Precautionary Ecosystem Health Principle, Mark Ryan
Normative Identity, Per Bauhn
Incommensurability and its Implications for Practical Reasoning, Ethics and Justice, Martijn Boot
Charles Taylor's Doctrine of Strong Evaluation: Ethics and Ontology in a Scientific Age, Michiel Meijer
On Music, Value and Utopia: Nostalgia for an Age yet to Come? Stan Erraught
Partial Values: A Comparative Study in the Limits of Objectivity, Kevin DeLapp
The Reality of Money: The Metaphysics of Financial Value, Eyja M. Brynjarsdóttir
Sovereignty as Value, Edited by André Santos Campos and Susana Cadilha
Stoic Philosophy and the Control Problem of AI Technology: Caught in the Web, Edward Spence
Normativity in African Regional Relations, Frank Aragbonfoh Abumere (forthcoming)
Ideals and Meaningfulness, André Grahle (forthcoming)
Psychosis, Refusal and Autonomy: A Phenomenological Study of Mental Health Detention, Owen Earnshaw (forthcoming)
Inhabiting Difference: Harnessing Lived Experience for Creative Social Change, James Abordo Ong (forthcoming)

Stoic Philosophy and the Control Problem of AI Technology

Caught in the Web

Edward Spence

ROWMAN & LITTLEFIELD
Lanham • Boulder • New York • London

Published by Rowman & Littlefield
An imprint of The Rowman & Littlefield Publishing Group, Inc.
4501 Forbes Boulevard, Suite 200, Lanham, Maryland 20706
www.rowman.com

Copyright © 2021 by Edward Spence.

All rights reserved. No part of this book may be reproduced in any form or by any electronic or mechanical means, including information storage and retrieval systems, without written permission from the publisher, except by a reviewer who may quote passages in a review.

British Library Cataloguing in Publication Information Available

Library of Congress Cataloging-in-Publication Data

Names: Spence, Edward, 1949- author.
Title: Stoic philosophy and the control problem of AI technology : caught in the Web / Edward Spence.
Description: Lanham : Rowman & Littlefield, [2021] | Series: Values and identities : crossing philosophical borders | Includes bibliographical references and index.
Identifiers: LCCN 2021035228 (print) | LCCN 2021035229 (ebook) |
  ISBN 9781786615916 (cloth) | ISBN 9781786615923 (epub)
Subjects: LCSH: Well-being. | Stoics. | Artificial intelligence—Social aspects. | Control (Psychology)
Classification: LCC BD435 .S64 2021 (print) | LCC BD435 (ebook) | DDC 128—dc23
LC record available at https://lccn.loc.gov/2021035228
LC ebook record available at https://lccn.loc.gov/2021035229

To Kaye

Contents

Acknowledgements ix

1 Introduction: Who Is in Control? 1
2 What Is Technology Good For? 11
3 Stoic and Neo-Stoic Philosophy 27
4 Application of Stoic Philosophy to Technology 59
5 Wisdom and Well-Being: The Dual Obligation Information-Wisdom Theory 73
6 Tech Media Corruption in the Age of Information 101
7 The Normative Impact of ICT Technologies on Well-Being 125
8 The Normative Impact of AI Technologies on Well-Being 149
9 Smart Machines and Wise Guys: Who Is in Control? 189

Index 233
About the Author 247

Acknowledgements

I would like to express my heartfelt thanks and gratitude to the following people who in various ways have helped me bring this book to fruition. The acquisition editor from Rowman and Littlefield, Tsarina Doyle, for her constant support throughout the publication process of this book. My mentor, Alan Gewirth, my friends and colleagues John Weckert, Seumas Miller, Rick Benitez, David Field, Chris McGillion, Ben Chegwin and Miles Downie from the Arena Gym, University of Sydney, Montana, and Richard Spence, for their useful comments and discussions on aspects of this book and all my colleagues from the *4 TU Centre for Ethics and Technology*, as well as my long-standing colleagues, Luciano Floridi, Charles Ess, Johnny Soraker, Adam Briggle, Jeroen van den Hoven, Judith Simmons, Peter Paul Verbeek. Ibo van de Poel, and Philip Brey, and not least Plato, Socrates, Zeno of Citium, Marcus Aurelius and especially the Sempiternal Muse, for their abiding presence and constant inspiration.

Chapter 1

Introduction

Who Is in Control?

> *The first ultraintelligent machine is the last invention that man need ever make, provided that the machine is docile enough to tell us how to keep it under control.*
>
> —Irving J. Good, 1965 (from Max Tegmark's book *Life 3.0*, 2018, 134).

RATIONALE AND APPROACH

The two most urgent global problems facing us today are climate change and the rapid technological advances in autonomous artificial intelligent agents. The first concerns the biosphere and the second just as significant, the infosphere. Both pose an existential risk that potentially if not resolved threaten the extinction of the human race.

This book is about the second existential problem facing society and the world community, and specifically the *Control Problem*: that is, who is in control of the information and communication technologies (ICTs) and artificial intelligent (AI) technologies? Is it Us collectively as the global community, or the autonomous ICT and AI technologies that presently are under the control of the Big Tech companies, such as Facebook, Google, Amazon, Microsoft and Apple, known as the Big Five?

What is technology good for? This book is motivated by the need to answer that question by examining the impact technology has on human well-being, and specifically for the purpose of this book technology in the form of ICTs, as well as its closely related AI technologies. For in the absence of any tangible or potential *eudaimonic* benefit for society on our collective well-being, what is technology good for, especially in cases where the eudaimonic impact of a particular technology on society is not only very negligible approaching zero but negative and harmful?

To answer that question, the book develops and applies a meta-normative model whose primary purpose is to determine the essential conditions that any normative theory that seeks to assess the impact of technology on well-being must adequately address to be able to account for, explain and evaluate its *Contributive Capability of Technology for Well-being* (CCT-WB) (Spence 2011). By CCT-WB, we understand the *capability* of any technological product or process in its design and/or its use to contribute in some way, if any, to the good life for the *well-being* of individuals and society generally. In this book, the all-embracing term *technology* will be used to refer to both ICT and AI technologies.

An answer to the question, what is technology good for, is also examined in chapter 3, *Stoic and Neo-Stoic Philosophy* as Stoic philosophy provides the overarching normative evaluative theory that informs this book and is applied in chapter 4, and throughout this book, to assess the normative impact of technology on well-being. Moreover, the application of Stoic philosophy to Technology is the unique and innovative feature of this book.

Using a theoretical normative tripartite model, the *Dual Obligation Information Theory-Wisdom (DOIT-Wisdom)* (Spence 2011) that comprises three interrelated normative components, the *epistemic*, the *ethical* and the *eudaimonic*, within the overarching conceptual framework of Stoic and neo-Stoic philosophy, the book examines the CCT-WB by identifying and evaluating any technology's actual or potential impact on well-being. For example, ICTs and AI technologies that do not protect the justified autonomy and privacy rights of its users or allow and do not restrict the indiscriminate dissemination of misinformation and disinformation, as well as the indiscriminate harvesting of user's information for the primary gain of Big Tech companies such as Facebook and Google, would have a low and negative CCT-WB as they would potentially generate a negative epistemic, ethical and eudaimonic impact for its users and generally, a negative impact on society and the global community overall.

The main argument of the *DOIT-Wisdom* is that the notion of *wisdom* understood as being at once a *meta-epistemic*, *meta-ethical* and *meta-eudemonic* concept, provides the essential conceptual link between information on the one hand, and the good life for the attainment of eudaimonia or well-being, on the other. In the digital age of information, both the theoretical examination and analysis of the question of how information in its technological construction and dissemination relates to the good life and the attainment of well-being, as well as the provision of an adequate answer to that question, are both essential for developing a deeper understanding of how to evaluate the theoretical and practical implications of the impact of ICT and AI technologies on well-being, for individuals and societies generally.

MAIN THEMES AND OBJECTIVES

The book focuses mainly on examining the CCT-WB of ICT and AI technologies as they are two of the most prominent, ubiquitous and impactful types of technologies at present with the potential of the greatest good or the greatest harm for humanity. By normatively assessing the CCT-WB of those two general types of technologies, this book seeks to identify their actual or potential epistemic, ethical and eudaimonic impact on society through the application of the DOIT-Wisdom model applied to some paradigmatic case studies with regard to the core and general normative issues raised by those technologies. The use of some key case studies helps identify, illustrate as well as contextualise the problematic normative issues of those technologies.

In the case of *ICT technologies,* the key normative issues examined in this book are *privacy, transparency, accountability, opacity, truth and trust,* as well as the *corruption of information* (Spence 2021, 2017). The scandal of Cambridge Analytical and Facebook involving breach of privacy and lack of consent by its users, as well as the unlawful electoral interference with the Brexit Referendum in the UK and the presidential elections in the United States of 2016, is a case in point involving not only the breach of privacy but also the dissemination of misinformation and disinformation that has come to be infamously known as *fake news*, *alternative facts* and *post-truth*. This has resulted in mistrust of the media and mistrust in its ability to communicate truthful information to the public on matters of public interest, which, in turn, has caused mistrust in government, including political, judicial and media institutions of the 4th Estate. The outcome is that it has undermined democracy itself. In this particular case and other similar cases like it, for example, the previous *News of the World* scandal in the UK, it is clear that all three normative aspects, the epistemic, the ethical and the eudaimonic were undermined and resulted in a negative impact on our individual and collective well-being as a society.

In the case of AI technologies, a string of successes including IBM's *Deep Blue* that beat the World Chess Champion Kasparov in 1997, IBM's *Watson* that beat the World champions in the game Jeopardy! in 2011, and most recently DeepMind's *AlphaGo* that beat the World champion in the game Go in 2016, has made rapid advances through big breakthroughs in machine learning, deep learning and neural networks. This has begun a worldwide debate on the benefits and risks involving the use of highly intelligent autonomous AI agents across different applications, such as driverless cars, autonomous weapons, applications in medicine and the health service, and most importantly, the potential existential risk that *Superintelligent AI agents* pose for humanity (Nick Bostrom 2015, Max Tegmark 2018, Stuart Russell 2019, as well as concerns expressed by Elon Musk and the late Stephen Hawking, among many others AI researchers).

As in the case of ICT technologies discussed earlier, the CCT-WB of AI technologies will be assessed through the application of the DOIT-Wisdom model, under the overarching critical perspective of Stoic philosophy, to identify and evaluate some of the core epistemic, ethical and eudaimonic issues of AI technologies, and their impact on the well-being of society and the global community. Privacy, accuracy, transparency, opacity, autonomy, accountability and consent, as in the case of ICT technologies, remain also problematic issues in the case of AI technologies that those technologies seem to exacerbate. For example, the use of facial recognition and profiling through AI technologies has in some cases resulted in police misidentification of groups of people with regard to inaccurate racial profiling that produces a higher percentage of African Americans suspected of crime in the United States. Again, the application of the DOIT-Wisdom model to this particular type of normative issue illustrates that all three normative principles the epistemic, ethical and eudaimonic are negatively impacted on, through the indiscriminate and inaccurate use of those technologies.

Apart from these general types of applied normative issues with AI technologies, the most problematic are the dual interrelated problems of *inscrutability* and *control* that raise a metaphysical and existential concern about our ability to control such highly intelligent autonomous AI agents that could, in the not-so-distant future, outpace us in intelligence and place us under their control with a potential risk to our own existence (Bostrom 2015, Tegmark 2018, Russell 2019).

The present problem with the inscrutability of AI systems that produce results through the use of machine learning and neural networks is that the processes of arriving at those results are not transparent and well-understood by the AI scientists themselves, least of all by the end-users of those results and the impact they might have on society presently and in the future. The inscrutability problem, in turn, relates to the control problem. For the opacity and lack of transparency of these AI systems render those systems potentially uncontrollable and in the not unlikely possibility that in the case of a sudden and unpredictable explosion of AI Superintelligence that overtakes any attempt on our part to control such systems, AI might pose a serious existential risk to the survival of human race. Again, we can see that this metaphysical AI problem is subject to a normative analysis using the DOIT-Wisdom model that can identify and *anticipate* in advance the potential impact on society: the epistemic (lack of transparency and epistemic opaqueness and unknowability); the ethical (harm to human beings) and the eudaimonic impact (major negative impact on societal well-being, potentially leading to humanity's extinction).

Apart from normative theoretical completeness, another methodological advantage of using the tripartite DOIT-Wisdom model is that it also provides an associative comparative risk analysis in an ascending normative hierarchy, the eudaimonic together with the ethical taking precedence over the epistemic. For given that technology cannot be good just for itself but good for humanity, then even if AI applications produce more instrumental and epistemic value but at the cost of higher ethical and eudaimonic negative impact on society, then all things being equal, ethical and eudaimonic considerations should always take precedence over instrumental and epistemic considerations. By way of an illustrative example, AI applications using neural networks produce results with a much higher epistemic accuracy but also a higher level of inscrutability, which could potentially also produce adverse ethical and eudaimonic outcomes for society if the rationale for arriving at those results remains hidden and non-transparent to human agency and oversight.

Another central and overarching theme of this book, as previously mentioned, is the use and application of *Stoic and neo-Stoic philosophy* (collectively referred to as Stoic philosophy) in analysing and evaluating the two main themes of ICT and AI technologies referred to above. The rationale for this approach is that Stoic philosophy, in particular *Stoic ethics*, with its central focus on the attainment of eudaimonia through virtue is singularly relevant to the eudaimonic assessment of ICT and AI technologies to be conducted in this book. For the DOIT-Wisdom model, which is used in this book to evaluate the epistemic, ethical and eudaimonic impact of technology on well-being is by design both theoretically and practically akin and closely related to the operationalisation and application of Stoic philosophy, through its alignment of rationalist and virtue ethics (Spence, E. 2016 and Spence, E. 2006).

One central aspect of Stoic philosophy that is methodologically explored in this book is the issue of *Control*, which as explained earlier is the core problem that arises in both ICT and AI technologies. For the notion of control is also a central feature of Stoic philosophy, and therefore conceptually and methodologically, well placed in providing useful conceptual analysis and evaluation of the control problem at it manifests itself in the two relevant technologies, ICT and AI technologies, that are the main focus of this book.

According to Epictetus,

> Some things are up to us and others are not. Up to us are opinion, impulse, desire, eversion, and in a word, whatever is our own action. Not up to us are

body, property, reputation, office, and, in a word, whatever is not our own action. (*The Handbook of Epictetus*, Chapter 1)

In summary, according to the Stoics, we should only be concerned with developing our virtues (courage, moderation, justice and prudence or practical wisdom) as the only true good in itself, which is both necessary and sufficient for human eudaimonia. If core aspects of technology are outside our control such as the ones described above, there are at least two options open to us within the Stoic ethical methodology. Option one, is that since accuracy, transparency, privacy, the inscrutability and ultimately control of AI autonomous systems are beyond our control we should ignore them or at least treat them as preferred indifferents not required for our well-being or eudaimonia since we can't control them; Option two, is that precisely because of the potential harmful impact those technologies can have on the epistemic, ethical and eudaimonic aspects of our lives, we should in fact endeavour to bring those technologies under our human societal control, since our societal well-being is at risk if we don't.

Given that *eudaimonia* or *well-being* is the primary concern of Stoic philosophy, it is reasonable and prudent to adopt the second option of bringing those presently uncontrollable aspects of technology within our societal control since our collective well-being is at risk if we don't. And that is the option that this book pursues, which also is of a piece with the proposed application of the DOIT-Wisdom model in evaluating in detail through the use of case studies the key normative issues that arise in ICT and AI technologies.

CHAPTER SYNOPSIS

Chapter 2: What Is Technology Good For?

An overview of some key case studies that illustrate the positive and negative impacts on well-being of ICT and AI technologies to prime and set up the motivation for the necessity of evaluating the impact of technology in terms of well-being, since technology is not of itself valuable unless it benefits human society.

Chapter 3: Stoic and Neo-Stoic Philosophy

This chapter introduces Stoic philosophy as the overarching and unique feature of this book that guides the normative evaluation of the impact of technology on well-being. The chapter also introduces rationalist and virtue ethics from a neo-Stoic perspective as an essential component to the enquiry of the book around the notion of well-being and eudaimonia (Spence, E. 2016; Spence, E. 2006).

Chapter 4: Application of Stoic Philosophy to Technology

This chapter examines how Stoic and neo-Stoic philosophy applies to technology, and specifically ICTs and AI technologies. Unless otherwise stated for convenience, I will refer to these technologies collectively, as "technology." The chapter comprises three parts: (1) The core problem of Stoic philosophy and technology; (2) The core principles of Stoic philosophy and how they apply to technology and (3) Why Stoic philosophy is of relevance and importance to the normative evaluation of technology.

Chapter 5: Wisdom and Well-being: The Dual Obligation Information-Wisdom Theory

This chapter explains in detail the rationale and the application of the DOIT-Wisdom model in normatively evaluating the impact of ICT and AI technologies in terms of the model's three interrelated components the epistemic, ethical and eudaimonic (Elliott and Spence 2018; Spence 2011 and 2009). A central aspect of this chapter is the concept of *wisdom*, as a triadic concept comprising epistemic, ethical and eudaimonic values, that conceptually and normatively links information to well-being.

Chapter 6: Tech Media Corruption in the Age of Information

This chapter examines how some of the practices of Facebook and Google constitute systemic *Tech Media and Information Corruption* (Spence, Edward H. 2020, 2017) that undermines both societal well-being and democratic institutions.

Chapter 7: The Normative Impact of ICT Technologies on Well-Being

This chapter examines and evaluates in detail the normative impact of ICT technologies such as Facebook and Google on well-being. Issues examined are privacy, transparency, accuracy, truth and trust and the illustration and contextualisation of those issues through key case studies (Spence, E. 2018; Spence, E. 2011).

Chapter 8: The Normative Impact of AI Technologies on Well-Being

This chapter examines and evaluates the impact of AI technologies on well-being through a close analysis of two major design problems in AI machine learning and neural network algorithms: that of the *inscrutability* or *black box* problem and the *control problem*. The chapter also suggests possible solutions to both of those problems, some of them proposed recently by AI researchers. A key concern that both those problems raise is that without transparency there is no accountability and without accountability there is no normative way to monitor and control AI systems for potential adverse and harmful impacts on society. The chapter also introduces the notion of anticipatory normative controls to prevent or at least minimise the risk of adverse impacts of AI on well-being, through opacity or lack of control.

Chapter 9: Smart Machines and Wise Guys: Who Is in Control?

This concluding chapter examines and argues for the adoption of a *Eudaimonic Stoic Stance* and *Humanistic Approach* towards the design and use of ICT and AI technologies in particular, with the aim of (a) preventing or at least minimising the negative and harmful impacts of technology on human well-being and (b) enhancing the design and use of technology for the overall eudaimonic benefit of humanity presently and in the future.

REFERENCES

Bostrom, Nick. 2015. *Superintelligence: Paths, Dangers, Strategies.* Oxford: Oxford University Press.

Brey, Philip, Briggle, Adam, and Spence, Edward H. (eds.) 2012. *The Good Life in a Technological Age.* New York: Routledge.

Elliott, Deni and Spence, Edward H. 2018 *Ethics for a Digital Era.* Oxford: Wiley-Blackwell.

Epictetus. 2018. *Epictetus: Discourses and Selected Writings,* translated and edited by Robert Dobbin. London: Penguin Classics.

Hui, J. and Spence, Edward H. 2016. Internet Addiction and Well-Being: Daoist and Stoic Reflections. *Dao: A Journal of Contemporary Philosophy* 15:209–225. DOI 10.1007/s11712-016-9488-8. Switzerland, AG: Springer Nature.

Russell, Stuart. 2019. *Human Compatible: AI and the Problem of Control.* UK: Penguin Books.

Spence, Edward H. 2021. *Media Corruption in the Age of Information.* Switzerland, AG: Springer Nature.

Spence, Edward H. 2017. Corruption in the Media. In Michael S. Aßländer and Sarah Hudson (Eds.), *The Handbook of Business and Corruption: Cross-Sectoral Experiences.* UK: Emerald Publishing.

Spence, Edward H. 2016. Is Alan Gewirth's Moral Philosophy Neo–Stoic? In Per Bauhn (Ed.) *Gewirthian Perspectives on Human Rights.* New York: Routledge.

Spence, Edward H. 2013. "Wisdom and Well-being in a Technological Age". *Synesis: A Journal of Science, Technology, Ethics, and Policy.* Arlington, VA: Potomac Institute Press.

Spence, Edward H. 2011. Is Technology Good for Us? A Eudaimonic Meta-Model for Evaluating the Contributive Competence of Technologies for a Good Life. *Nanoethics,* 5:335–343.

Spence, Edward H. 2006. *Ethics Within Reason: A Neo-Gewithian Approach* (2006). Lanham, Maryland: Lexington Books (a division of Rowman and Littlefield, USA).

Tegmark, Max. 2018. *Life 3.0*: *Bring Human in the age of Artificial Intelligence.* UK: Penguin Books.

Chapter 2

What Is Technology Good For?

What is human flourishing, or "the good life," and how can technology contribute to this?

—John Weckert, 2013.

INTRODUCTION

The title refers to the question addressed in this chapter, namely, to what degree if any technology, in the form of products and processes, is capable of contributing to a good life. To answer that question, the chapter will develop a meta-normative model whose primary purpose is to determine the essential conditions that any normative theory of the good life and technology (T-GLAT) must adequately address in order to be able to account for, explain and evaluate the contributive capability of technology for a good life (CCT-GL). By CCT-GL, we understand the capability of any technological product or process in its design and/or its use to contribute in some way, if any, to the good life of individuals and society at large. In this chapter, the all-embracing term "technology" will be used to refer to both the products and processes of different technologies.

In an entry in the *International Encyclopedia of Ethics*, "Nanoethics," John Weckert (2013) writes that

> Nanotechnologies . . . involves consideration of what ought to be researched and developed and whether the precautionary principle should be applied in some areas, and this in turn encourages examination, or re-examination, of some basic issues in the ethics and philosophy of technology and science: What

is the purpose of technology? Can technological research and development be directed externally, and if it can be, should it be, and by whom? What is human flourishing, or "the good life," and how can technology contribute to this? These questions are larger than nanoethics but if the ethics of nanotechnology are to be discussed usefully, they cannot be separated from those discussions.

I concur fully with that assessment and its sentiment concerning not only nanotechnologies but all technologies, including emerging technologies more generally, and specifically, *Information and Communication Technologies (ICTs) and Artificial Intelligence* (AI) technologies, which is the main focus of this book. The primary aim of this chapter is to address a general question in the philosophy and ethics of technology, namely, to what degree if any, technology, in the form of products and processes, is competent in contributing to a good life for the attainment of well-being. An answer to that question is also examined in chapter 3, *Stoic and Neo-Stoic Philosophy*. To answer that question, this chapter will utilise a meta-normative model whose primary purpose is to determine the essential conditions that any normative T-GLAT must adequately address to be able to account for, explain and evaluate the CCT-GL.[1] By CCT-GL, we understand the capability or capacity of any technological product or process in its design and/or its use to contribute in some way, if any, to the good life of individuals and society, and humanity at large, for the attainment of well-being. In this chapter, the all-embracing term "technology" will be used to refer to both the products and processes of different technologies.

TWO META-CONDITIONS FOR ANY NORMATIVE T-GLAT

There are at least two necessary methodological *meta-conditions* that any normative T-GLAT must meet. I will refer to those conditions as the *formal condition* and the *material condition*.

The Formal Meta-Condition

The formal meta-condition characterises the necessary *structural form* of a T-GLAT's theoretical framework. It comprises at least three general normative categories that any T-GLAT must of necessity include within its theoretical framework, so as to be both *theoretically and practically adequate*. Those formal normative categories are *motivation, justification* and *compliance*. For if it lacks justification, rational agents will have no reason to be convinced of its rational authority and if it lacks sufficient motivation, agents will not be pre-disposed to act in accordance with its prescriptions. Therefore, rational agents shall have no reason to offer their rational allegiance to a T-GLAT that lacks rational cogency. Finally, compliance is *practically necessary* if the prescriptions of a T-GLAT are to be capable of leading to social and political action through T-GLAT informed policies. In summary, the three formal meta-conditions are:

1. **Justification** – that which is adequately provided by rational arguments, capable of persuading and convincing any putative rational agent that they have good rational reasons or grounds for accepting and acknowledging the rational authority of certain precepts, rules, principles and values.
2. **Motivation** – that which motivates any putative rational autonomous agent to accept and acknowledge certain precepts, rules, principles and values, and moreover, would be adequately motivated to act in accordance with them.
3. **Compliance** – both motivation and justification are necessary for actual compliance with any T-GLAT so that it results in action or practice, but are not sufficient. This is because of at least two reasons: first, weakness of the will of individual agents (*individual akrasia*) or weakness of the will of a collective or group of individuals (*social akrasia*); and second, lack of political will in the form of the inadequacy or lack of appropriate policies that encourage and promote compliance with adequate policies, informed and supported by any T-GLAT, which otherwise meets the motivation and justification conditions.

The Material Meta-Condition

The material condition characterises the necessary *content* that any T-GLAT must address and include, both in its conceptual explanation and practical application. It comprises at least eight essential minimal conditions that any T-GLAT that seeks to evaluate the CCT-GL must adequately address:

1. Capacity to address *desires*
2. Capacity to meet *needs*
3. Capacity to provide *means-ends satisfaction* (instrumental satisfaction)
4. Capacity to contribute to *valuable-ends*
5. Capacity to preserve and promote *moral rights*
6. Capacity to be *sustainable*
7. Capacity to lead to the attainment of *well-being or eudaimonia*
8. Capacity to be *practical* in providing the ground for formulating policies in enabling compliance through its implementation and application.

These material conditions will be explained in more detail in the following.

The Two Theses of the Chapter in Summary

The above schema shows in outline how the formal conditions of justification, motivation and compliance are expressed and closely inter-linked contextually with the eight material conditions of the meta-model proposed. Using this meta-model, this chapter will argue for and support two inter-related theses:

(A) Any T-GLAT, whether desire-satisfaction theory, objectivist theory or capability theory, amongst others, must be capable of at least addressing and accounting for the two formal and material meta-conditions outlined above. Insofar as any T-GLAT fails to do so, it is not an adequate theory.
(B) An adequate T-GLAT theory to be proposed in this chapter is the *Eudaimonic Model for Evaluating the Goodness of Technology* (EMEGOT) based on the notion of wisdom (Spence 2011). The chapter will provide arguments to demonstrate and support the case for such a theory but without excluding the possibility that other theories that meet the necessary formal and material conditions outlined above might also prove successful for that purpose. As such, the chapter takes a *pluralistic methodological approach* to the research question addressed in this chapter. In chapters 3 and 4, we shall examine how and why Stoic and neo-Stoic philosophy provides an adequate T-GLAT theory for the eudaimonic evaluation of the goodness of technology for society and global community generally.

THE ADEQUACY OF THEORIES OF THE GOOD LIFE AND TECHNOLOGY

Theories of the good life and technology (T-GLAT theories) irrespective of their specific content are collectively defined in this paper as theories whose primary purpose is to demonstrate how technologies (any technology) contribute to a good life. These theories seek to address the theoretical question posed in this chapter, namely, whether and in what manner technologies are good for us, as individuals and societies, and more generally humanity, in enabling us to live good lives for the attainment of well-being or eudaimonia. A basic theoretical assumption in this chapter and generally in this book is that insofar as technologies are designed and used for the purpose of improving our lives then they must in some way contribute to the goodness of our lives. That is, technologies must, at least in some minimal sense, make a contribution in enabling us as individuals and as a society to live a good life. For if they made no such contribution, what would they be good for? Such technologies would be at best practically useless or at worst, bad. However, given the aforementioned material condition of sustainability, even a technology that was not directly bad for us as individuals, could still be viewed as collectively bad in some minimal sense if it were shown not to be sustainable.

What Is a Good Life?

We can say quite reasonably that a *good life* generally is one that is at least minimally capable of enabling a person to attain self-fulfillment,[2] well-being, happiness. I will for methodological convenience use the general term *eudaimonia*[3] to include and refer to all those concepts collectively whilst maintaining the original intended meaning for that term by the ancient Greek philosophers, including Plato, Aristotle, and the Hellenistic philosophers and in particular the Stoics and the Epicureans (Spence, 2006, Ch. 10). Although those philosophers might have explained the notion and attainment of eudaimonia in different ways, they all at least agreed that the virtues were essential for the attainment of a eudaimonic or a flourishing life and moreover, the virtues were constitutive of such a life. For insofar as eudaimonia is our ultimate object in life as Aristotle claimed, it is difficult to conceive a life that was not at least capable of leading to the attainment of eudaimonia, as good – what would it be good for if it were incapable of at least in principle enabling one to realise one's ultimate objective in life? In this chapter, I will therefore define a *goof life* as any life that is demonstrably capable of contributing to the attainment of eudaimonia (the collective term for well-being, happiness and self-fulfillment).

Formal and Material Conditions for Adequacy

Formal Conditions

Any T-GLAT to be adequate must satisfy as mentioned earlier both a formal and a material condition. To be formally adequate, it must be rationally justified, motivating and practical in the sense of being capable of resulting in compliance. Justification generally in decision-making is intended to provide convincing if not conclusive rational reasons for selecting one course of action rather than another and in the case of T-GLAT, rational reasons for selecting a particular T-GLAT rather than another. Justified reasons alone, however, are not sufficient to guide rational action. That is because rational action requires that the rational reasons of justification must also be motivating reasons. For if justificatory reasons are not motivating, they can't be practical and if they can't be practical, they are not action guiding and hence cannot play a role in the decision-making process. Thus, the decision-making process requires reasons for action that are at once justificatory and motivating. A crucial point with regard to motivation, however, is that justificatory reasons need only be *capable* of motivating action. They need not and cannot also be expected to motivate action in every single instance. Although capable of motivating action in normal rational agents, motivating justificatory reasons may nevertheless fail in some instances to motivate particular agents to act. In sum, justificatory reasons will count as motivating and thus practical and action guiding if they are *capable* of motivating a normal rational person to act in a certain way, specifically, for the purposes of this chapter, acting in giving their rational ascent to a particular T-GLAT.

In addition, any T-GLAT must be adequately practical in enabling compliance through informing policies that can promote its implementation. For a theoretically adequate T-GLAT that met both the justification and motivation formal conditions but was impractical in its application would not succeed in meeting the formal compliance condition. It would thus prove insufficient in satisfying its material conditions as specified above.

Material Conditions

To reiterate, the identified material conditions that any T-GLAT must address and satisfy for adequacy are the following:

1. Capacity to address *desires.*
2. Capacity to meet *needs.*
3. Capacity to provide *means-ends satisfaction.*
4. Capacity to contribute to *valuable ends.*
5. Capacity to preserve *moral rights.*
6. Capacity to be *sustainable.*
7. Capacity to lead to the attainment of *eudaimonia.*
8. Capacity to be *practical* in providing the ground for formulating policies in enabling compliance through its implementation and application.

Roughly, conditions 1–4 primarily provide the context and drive *motivation*, and 4–6 primarily provide the context and drive *justification*. Conditions 7 and 8 provide at once the context for motivation and justification and condition 8 carries most of the weight for providing adequate compliance in terms of the practical implementation of a T-GLAT. Thus, to be capable of motivation (a necessary formal condition), any T-GLAT must be capable of evaluating if and to what degree any technology addresses and accounts for the desires, needs, mean-end satisfaction (or instrumental strategic thinking) and the valuable ends (how valuable are the designated ends or goals promised or afforded by any technology) of actual agents.

In addition, any T-GLAT must be capable of providing adequate justification (also a necessary formal condition) of why such motivating states are reasonable on the basis of good reasons supporting those motivating states. Moreover, any T-GLAT must also be capable of accounting for and providing adequate justification of how and to what degree any technology preserves the basic minimal moral rights owed to every individual person, how and to what degree any technology is sustainable, and finally, how and to what degree any technology is capable of contributing to our individual and collective eudaimonia as a society.

THE EUDAIMONIC MODEL FOR EVALUATING THE GOODNESS OF TECHNOLOGY (EMEGOT)

In this section, I will propose and defend a Eudaimonic Model for evaluating the contribution that technology makes to the good life (its eudaimonic goodness). In asking the question, "what is technology good for," we can begin by saying that technology has generally some instrumental goodness as a means to attaining some functional goal or purpose. The instrumental goodness of an aeroplane, for example, lies in its capacity to transport passengers across the globe in less time than any other available means of commercial transportation, such as boats and trains.

Intrinsic but Conditional Goodness

Technological products and artefacts have some minimal intrinsic value and goodness by virtue of their designed-in-agentive-purposiveness. It is conditional on evolving human values and needs but technologies have no unconditional value or goodness in themselves (Spence, 2010).

Technology has minimal intrinsic but conditional value only to the extent that it contributes to meeting some specified functional human goals or purposes. So insofar as technology is not good *simpliciter*, it is good only to the extent that it has the capacity to contribute to the human good. People have a vast array of different instrumental goals to which technology can contribute as a means of achieving those goals. Is there one ultimate goal that all people value and desire as an end in itself? It is reasonable to assume that most if not all people aspire to have a good life capable of contributing to the attainment of eudaimonia (self-fulfillment, well-being, happiness, flourishing). This is in keeping with the *Eudaimonist axiom*, the view that "happiness is desired by all human beings as the ultimate end or telos of all rational action" (Brink 1999, 255).[4]

Undoubtedly, technology contributes in countless ways to the good life instrumentally in meeting evolving needs, desires and valued individual and collective ends such as transportation, health, wealth, power, education, communication, and so on. How can we normatively evaluate technology's variable instrumental contributions to the good life? In what follows, I will argue that the answer lies in technology's capability to contribute to the attainment of a good life. What value can we use to normatively evaluate technology's capability in contributing to a good life (CCT-GL)?

Wisdom as the Principle for Assessing Technologies for a Good Life

Insofar as the ultimate goal of a good life is the attainment of eudaimonia, we can evaluate a technology's capability of contributing to a good life (CCT-GL) by ascertaining its capacity for contributing to a good life for the attainment of eudaimonia. In this chapter and throughout this book, I will refer to technology's capacity to contribute to a good life as its *eudaimonic value*. A technology's eudaimonic value therefore relates and is directly proportional to its capacity to contribute to a good life, capable of leading to the attainment of eudaimonia. In this chapter, I shall demonstrate that a technology's eudaimonic value can be determined directly by the application of a model based on the notion of Wisdom (Spence 2011).

In the first instance, I define *wisdom* as a type of *meta-knowledge* and an enabling *second-order reflective virtue* whose application is capable of guiding one in conceiving and discovering what a good life is and applying that knowledge in its active pursuit for the attainment of eudaimonia. Wisdom provides the overall answer to the question of why we need to design and use technologies in general: because they are capable of contributing to human eudaimonia. In addition, wisdom enables us to evaluate the capacity of specific technologies to contribute to a good life for the attainment of eudaimonia.

Wisdom as a type of meta-knowledge provides *why-answers* – why design and use certain technologies in the first place; and as an enabling meta-axiological-virtue for conceiving what a good life is for the attainment of eudaimonia, wisdom provides *how-answers* of how to use those technologies in pursuit of a good life for the attainment of eudaimonia. Therefore, wisdom provides a theoretical and practical model for evaluating why and how certain technologies are good for us by ascertaining their capability for contributing to a good life for the attainment of eudaimonia. This is essentially the core argument of this chapter.

In what follows, I shall examine more closely how technology (understood as a collective term for all technologies) can be directly related to the notion of a good life via the concept of wisdom. Insofar as wisdom is a primary and essential condition for an individual in (a) determining what a good life is or ought to be (meta-knowledge-that and meta-knowledge-why); (b) a primary and essential condition in providing us with guidance and direction, both as individuals and societies generally, of how to live such good lives; and (c) wisdom, as a reflective meta-virtue, is a disposition of character that practically enables us to live such good lives for the attainment of eudaimonia (meta-knowledge-how); to what extent and in what ways, if any, can wisdom provide guidance in identifying and evaluating the degree by which technology contributes to the good life for the attainment of eudaimonia?

This chapter posits that one direct way of evaluating the value of technology and its capacity to contribute to a good life generally (its overall axiological goodness) is by determining the degree to which it contributes or is capable of contributing to the attainment of a good life: *epistemologically* (its capacity to yield knowledge); *ethically* (its ability to contribute to the moral good of oneself and others both negatively by not causing unjustified harm to others, and positively by causing positive good for oneself and others); and *eudaimonically* (its capacity to contribute to both the conception and the attainment of a good life for the attainment of eudaimonia). This chapter will show that to achieve that theoretical objective the notion of wisdom[5] is essential. In sum, insofar as the ultimate purpose of a good life is the attainment of eudaimonia then wisdom, which informs the conception of a good life and directs its active pursuit for the attainment of eudaimonia, is an essential condition for both the conception and the attainment of a good life.

As the essential condition for both the conception and guided active pursuit and successful achievement of a good life, wisdom is therefore established as an essential conceptual link between technology and the good life, and in particular for evaluating the eudaimonic contribution that various technologies make to a good life. This, in turn, allows us to determine some of the generic implications and ramifications of technology for the conception of a good life, in particular, a eudaimonic conception of a good life. However, as Kekes points out,

> the eudaimonic conception of a good life is not to be understood as the endorsement of a particular form of life. It is rather a *regulative ideal* that specifies some general conditions to which all good lives must conform [emphasis added]. (Kekes 1995, 24)

The eudaimonic account of a good life canvassed in this chapter is broadly speaking pluralistic as it is in principle compatible with other different conceptions of a good life that meet the same necessary general conditions to which any notion of a good life must conform. For example, insofar as hedonistic, desire-satisfaction and objective list theories of the good life meet the minimal conditions for both specifying what a good life is as well as providing the enabling conditions for its practical realisation, then they too can be aligned broadly to the notion of wisdom developed in this chapter. To the extent that they meet those conditions, they too can be used to determine the contributive capability of technologies to a good life.

Wisdom as Meta-Technology of the Self

Insofar as technologies with a positive eudaimonic value are contributive, instrumental means to the end or *telos* for having a good life, and wisdom is the meta-knowledge (second-order knowledge) for providing the conception of a good life and guiding its realisation, wisdom can be considered as a *meta-technology of the self*. For it provides both the theoretical and practical *means* for the conception, design and realisation of a good life for the *end* for attaining eudaimonia. The means to the realisation of a good life may include and often does include first-order technologies such as computers, ICTs and AI Technologies, for example. If computers can be considered as extensions of the self, then the degree by which they form part of the self also becomes a question concerning wisdom.[6] How and to what degree computers and other technologies, as extension of the self, contribute to a good life for the attainment of eudaimonia.

To the extent that first-order technologies provide the means for making our lives better by contributing to a good life and wisdom provides the means as a *meta-technology of the self* for enabling human beings to have a good life for the attainment of eudaimonia, then clearly wisdom as a meta-technology should direct the choice and design of first-order technologies; for those will be the technologies that will have the highest eudaimonic value and the highest capacity for making a positive contribution to a good life.

A SUSTAINABLE GOOD LIFE AND TECHNOLOGY

A good life in the twenty-first century and beyond, however, should be a sustainable good life.[7] In the past, the issue of sustainability and that of a good life could have been perceived as two conceptually distinct and practical issues that could have been dealt with independently of each other. However, that is no longer the case. Since at least the Kyoto Protocol, the problem of sustainability has been the central focus of social, scientific and political debate. As examples of sustainable technologies, consider the dual-flush toilet and shower heads that are designed for saving water, recycling of water, recycling of consumables, such as plastic, glass and paper, installation of electrical household appliances, solar panels, as well as the design and production of transport vehicles including cars, buses, trains and aeroplanes that are more energy efficient and environmentally sound (the new *Dreamliner* aeroplane made entirely of carbon fibre for more energy efficiency is a case in point).

In all these examples, what seems crucial, however, is the role that people's desires play in motivating *compliance* with energy-saving technologies. People must want to comply and moreover be persuaded that there are justified reasons for complying with such technological energy-saving policies and practices for the policies and practices to work. The installation of an energy-consumption metre inside one's home, for example, that allows the monitoring of the overall energy consumption of a household would be a practical technological device for encouraging compliance with efficient energy use. Crucially, however, individuals must first be motivated to want to install such a device within their homes. Such initial motivation can best be primed by rationally targeting the consumers' desires (Spence, 2008).

The EMEGOT, which relates the goodness of technology to its eudaimonic value through its capability to contribute to a good life, is able to motivate people's rational desires for a good life. It does so through its ability to demonstrate that, although life is not indefinitely sustainable, an unsustainable life for a projected defined period cannot be good as it is potentially self-defeating and incapable of leading to eudaimonia. By creating its own unsustainable desires, by an ever-increasing plethora of technological devices that consume rather than preserve scarce resources and are thus non-sustainable in a technological world of ever-diminishing resources, it would be wise to design and make technologies that are compliant to the conditions for a sustainable good life. This could be accomplished by designing technologies that are capable of contributing to a sustainable good life for the attainment of eudaimonia.

We need to shift the focus from the epistemic value of technologies, their ability to contribute to knowledge or information (as in the case of communication and information technologies) for knowledge's sake, to that of their eudaimonic value, their ability to contribute to a sustainable good life capable or leading to eudaimonia. To do that, we need to design technologies wisely and that, in turn, requires that we enable ourselves to become wiser through designing and promoting *meta-technologies of the self* that enhance our capacity for wisdom. Revisiting and revitalising the ancient philosophies of Plato, Aristotle, the Epicureans, the Stoics and even the Pyrrhonic Sceptics, which were essentially practical philosophies of the good life for the attainment of eudaimonia, might be a good place to start. The primary object of this book, one examined in chapters 3 and 4, is to show how Stoic philosophy is fit for that purpose.

THE EUDAIMONIC MODEL FOR EVALUATING THE GOODNESS OF TECHNOLOGIES

The primary aim of this chapter has been to develop a meta-normative model whose primary purpose is to determine the essential conditions that any normative T-GLAT must adequately address to be able to account for, explain and evaluate the CCT-GL. I have argued in section, The Adequacy of Theories of the Good Life and Technology, that my prosed Eudaimonic Model, is an adequate normative model for evaluating the contribution that technology makes to the good life (its eudaimonic goodness) for it meets both the essential meta-formal as well as meta-material conditions for such an axiological evaluation. However, as I emphasised above, my choice of a Eudaimonic Model based on the concept of wisdom, does not of itself exclude other possible normative models that may also prove adequate. As aforementioned, the Eudaimonic Model described in this chapter will be applied to evaluate the capability of two specific technologies to contribute to a good life – that is, to evaluate their specific CCT-GL, and specifically, those of ICTs and AI Technologies. The reason for that choice is in part the great eudaimonic impact those technologies have had and continue to have on all aspects of society and its institutions, through the extensive social, political and financial influence and monopolistic power of the BigTech companies such as Facebook, Google, Amazon, Apple and Microsoft.

It is reasonable to assume pending further detailed examination that different technologies will have a variable CCT-GL – some higher, others lower – which will vary according to the specific function and use of those technologies. For example, the use of some communication and information technologies, such as the internet and smart-phones, operated by Facebook, Google and Apple, for example, may prove self-defeating for individuals and collectively for society and more generally humanity, by causing more harm than good, an issue that will be examined in detail in chapters 6–9. The Cambridge Analytical scandal involving Facebook, which influenced the U.S. Presidential Elections and the UK Brexit Referendum of 2016, is just one example that will be discussed in detail in chapter 6.

Other AI technologies such as automated predator drones and fully autonomous weapons used in war will require a more complex analysis to determine their CCT-GL. However, their capability for contributing to a good life should still be the overall guiding evaluative principle to be applied in normatively assessing those technologies. Determining the different general types of technologies and their corresponding CCT-GL according to their different functions and uses will make a worthwhile topic for further research.

CONCLUSION

The eudaimonic model for evaluating the goodness of technology (EMEGOT) proposed in this chapter is an adequate theory for evaluating the capability of technologies to contribute to a good life because it meets the essential formal and material conditions of an adequate T-GLAT.

First, it meets the formal conditions of justification, motivation and compliance: justification, because the three essential categories that comprise the notion of wisdom, the *epistemic*, the *ethical* and the *eudaimonic* are at once the essential normative categories required for evaluating the capability of technologies for contributing to a good life; motivation, because the notion of wisdom is capable of linking the normative evaluation of technologies directly to the notion of a good life and eudaimonia, which are highly valued practical objectives for most people – for most people wish to have a good life and to be happy; wisdom also as a *protreptic* value satisfies the compliance condition. For by being capable to directly link the contribution that technologies make to a good life, wisdom offers a practical measure for developing policies that can direct and guide the design of emerging technologies according to their eudaimonic value for the well-being of individuals, society and humanity generally.

Second, EMEGOT is also capable of meeting the material formal conditions of an adequate theory. It can address and account for *desires* and *needs* in terms of both their *instrumental means* and *final ends*, since wisdom relates those to the two highly valued aspirational objectives of a *good life* and *eudaimonia*. Finally, EMEGOT is capable of addressing and accounting for the material conditions of *moral rights* and *sustainability*, for wisdom as a meta-reflective-virtue also supports and promotes those essential material conditions. For as I have argued in this chapter, wisdom directs that a good life in the twenty-first century and beyond must of necessity be a sustainable life. We should, therefore, think and act more reflectively in examining the potential risks associated with technology and evaluate their potential benefits not primarily in terms of what new knowledge they give rise to but primarily in terms of the potential contribution they make to a good life (their eudaimonic value) for the overall benefit of individuals, society and humanity at large.

NOTES

1. This chapter is based on an earlier published version: Is Technology Good for Us? A Eudaimonic Meta-Model for Evaluating the Contributive Capability of Technologies for a Good Life. *Nanoethics* (2011a) 5: 335–343. DOI 10.1007/s11569-011-0134-y.

2. For an extensive discussion of self-fulfillment and how it relates to a good life see (Gewirth 1998). In his 2006 book (Chapter 10), Spence argues that Gewirth's theory of self-fulfillment is very similar in key aspects to Stoic eudaimonia. That topic will be examined in greater detail in chapter 3.

3. I will use the notions of eudaimonia and eudaimonic pluralistically as being potentially compatible with different theories of the good life, including hedonistic, desire-satisfaction and objective-list theories, among others. Simply put, as indicated earlier, the notion of a good life used in this paper is a good life that is in principle capable of leading to the attainment of eudaimonia. As such, any theory of a good life capable of leading to eudaimonia can at least in theory and upon further demonstration be considered a eudaimonic life. My own theoretical preference in this book is a eudaimonic life that includes the virtues in accordance with Stoic and neo-Stoic philosophy, a topic examined in chapters 3 and 4 but that need not exclude other theories capable of also leading to the attainment of eudaimonia (its overall axiological goodness). For a theory that takes an Aristotelian approach, see Shannon Vallor (2016) *Technology and the Virtues: A Philosophical Guide to a Future Worth Wanting*. Oxford: Oxford University Press.

4. Brink attributes the Eudaimonist axiom to Gregory Vlastos in Vlastos (1991) *Socrates: Ironist and Moral Philosopher*. Cambridge: Cambridge University Press. (Ithaca: Cornell University Press, p. 203).

5. For an extensive discussion of the concept of wisdom, see also Spence (2011).

6. For an influential paper on the theory of the Extended Mind, see Andy Clark and David J. Chalmers, *Analysis* (1998), 58:10–23.

7. For an extensive discussion of a sustainable good life, see Spence (2008).

REFERENCES

Brink, D. 1999. "Eudaimonism, Love and Friendship, and Political Community" *Social Philosophy & Policy* 16: 252–289.

Clark, Andy, and Chalmers, David J. 1998. *Analysis* 58: 10–23.

Gewirth, Alan.1998. *Self-fulfillment*. Princeton, NJ: Princeton University Press.

Kekes, J. 1995. *Moral Wisdom and Good Lives*. Ithaca: Cornell University Press.

Spence, Edward, H. 2011. "Information, Knowledge and Wisdom: Groundwork for the Normative Evaluation of Digital Information and Its Relation to the Good Life," *Ethics and Information Technology* 13(3): 261–275.

Spence, Edward H. 2011a. "Is Technology Good for Us? A Eudaimonic Meta-Model for Evaluating the Contributive Capability of Technologies for a Good Life. *Nanoethics* 5: 335–343. DOI 10.1007/s11569-011-0134-y.

Spence, Edward H. 2010. "Information Ethics Without Metaphysics: A Neo-Gewirthian Approach," *International Journal of Technology and Human Interaction* 6(1): 1–14.

Spence, Edward H. 2008. "Groundwork for a Neo-Epicurean Approach to a Sustainable Good Life in a Technological World," *Australian Journal of Professional and Applied Ethics*, 10(1 and 2): 73–81.

Spence, Edward, H. 2006. *Ethics within Reason: A Neo-Gewirthian Approach*. Lanham: Lexington Books (a division of Rowman and Littlefield).

Vallor, Shannon. 2016. *Technology and the Virtues: A Philosophical Guide to a Future Worth Wanting*. Oxford: Oxford University Press.

Vlastos, Gregory. 1991. *Socrates: Ironist and Moral Philosopher*. Ithaca: Cornell University Press.

Weckert, John. 2013. "Nanoethics," edited by La Follette, *International Encyclopedia of Ethics*, Wiley-Blackwell (accepted April 2011).

Chapter 3

Stoic and Neo-Stoic Philosophy

Here lies great Zeno, dear to Citium, who scaled high Olympus though he piled not Pelion on Osa, nor toiled at the labours of Heracles, but this was the path he found out to the stars – the way of temperance alone.

—Epitaph composed for Zeno, by
Antipater of Sidon, in Diogenes
Laertius, LCL, Book VII.

INTRODUCTION

In the last few years, there has been a notable and steady revival of Stoic philosophy. Scholarly works by A. A. Long, Gisela Striker, Julia Annas, Pierre Hadot, Martha Nussbaum, Brad Inwood, John Sellars and Laurence Becker,[1] to name but a few, have generated a renewed interest in Stoic philosophy and especially in Stoic ethics. Outside academic philosophy, Stoicism has also been exerting an indirect influence. I am referring to the remarkable rise of public philosophy, both offline and online, and practical and professional ethics, over the last twenty years or so. Professional ethics has been growing exponentially in almost all areas of professional practice, as in nursing, medicine, policing, engineering, computers, technology and the media. With regard to public philosophy, the dissemination and discussion of philosophical ideas in the context of issues relating to day-to-day living has also been taking place in public venues such as cafes, restaurants, theatres, pubs, vineyards, art galleries and even opera houses, across the world. Organised by professional and academic philosophers, these public philosophy forums have proved enormously popular with members of the general public, most of who have no prior philosophical background.[2] Generally, Hellenistic in spirit because of their practical orientation as a way of life, public philosophy, and to a lesser extent practical and professional ethics, have a Stoic pedigree, because of the robust professional, political, social and communal engagement that characterises them both.[3]

Unlike Epicureanism, Stoicism encourages active political and social engagement and not merely a retreat to a communal garden[4]; unlike Cynicism, Stoicism considers preferred indifferents like health, wealth and social status as desirable so long as they accord with a virtuous lifestyle. The Stoics, however, concur with the Cynics that only virtue is good and that it alone is both necessary and sufficient for a good and happy life, a eudaimonic life. Unlike the Sceptics, the Stoics do not suspend judgement on all issues concerning the natural, political and social environment they find themselves in. On the contrary, they counsel the active and rigorous pursuit of rational knowledge both of oneself and one's natural, social and political environments as a necessary condition for acquiring practical wisdom through virtue for the attainment of eudaimonia, the ultimate goal for the Stoics. Thus, the practical orientation, the vigorous engagement in the political, social and communal life of one's cultural and natural environments, both locally and globally, as well as the rigorous pursuit of rational knowledge as a necessary prerequisite in attaining practical wisdom essential for virtue and self-fulfillment, are central characteristics of Stoic philosophy and in particular Stoic ethics. In that regard, the proliferation of professional ethics and public philosophy in the pursuit of professional ethical knowledge and practical wisdom, even if not directly influenced by Stoicism is, at the very least, in terms of orientation and approach, attuned to some of its central precepts and principles. And that suffices to show that the practice of professional ethics and public philosophy is very much in keeping with the spirit and rationale of Stoicism, and more generally, Ancient Greek and Roman philosophy.

Stoicism also shares with other Hellenistic schools of philosophy, Epicureans, Cynics and Sceptics, the central notion that philosophy has to be practical; for it is only if it is practical that it can prove valuable in helping people to live good lives both as individuals and as societies for the attainment of eudaimonia or happiness. This notion is similar to one raised by Plato in *The Republic*, a notion that in all probability influenced the Stoics, namely that justice requires both personal and collective societal ethical harmony. It is perhaps that idea that informed and encouraged Stoics like Seneca and the Emperor Marcus Aurelius, to devote themselves to public life, although in the case of Seneca, as with Socrates before him, it cost him his life. There is always danger, fatal in Seneca's case, when a sage aligns himself to a serial killer. As Plato himself discovered with regard to the King of Syracuse, tyrants are not always amenable to the self-abstaining rigors of an ethical lifestyle.

The above preamble was intended primarily to indicate the tangible continuing legacy of Stoic philosophy and in particular ethics in contemporary practical and professional ethics, as well as in both scholarly and public philosophy. In what follows I will demonstrate that Alan Gewirth's rationalist ethical theory, in all its different facets, especially as set out in his book *Self-fulfillment* (1998), is paradigmatically and essentially Stoic in its structure, scope, rationale and vision, and schematically in its content. Moreover, I will show that Gewirth's ethical theory helps clarify and rationally support some of the central doctrines of Stoic ethics in a way that renders those doctrines more compelling and persuasive for a contemporary audience.

To demonstrate the essential Stoic features of Gewirth's ethical theory I will refer extensively to relevant passages in Gewirth's book, *Self-fulfillment* (1998), which bears close similarities to Stoic philosophy and ethics in particular, especially as regards the Stoic views on *eudaimonia* or well-being, virtue and reason. I will argue that according to my original interpretation and analysis, Gewirth's notion of *self-fulfillment* is a neo-Stoic version of Stoic *eudaimonia*.[5]

One key doctrine of Stoic ethics, which may first appear to be at odds with Gewirth's own views on the matter, is the controversial Stoic claim that virtue alone as the only good is both necessary and *sufficient* for happiness or eudaimonia (in this chapter, I will use those terms as well as that of well-being, interchangeably). At the outset, Gewirth appears to be more Aristotelian on this issue, who believed that though virtue was necessary for eudaimonia, it was not sufficient. However, there is a neo-Stoic way of understanding the necessity as well as the sufficiency of virtue for happiness, one that Laurence Becker convincingly expounds in his book *A New Stoicism* (1998), which, I believe, accords well with my analysis of the neo-Stoic view which I am ascribing to Gewirth in this chapter.

If I am right, the neo-Stoic character of Gewirth's ethical theory, to whom I am ascribing on the basis of my neo-Stoic reconstruction of his theory of *self-fulfillment*, is at once both surprising and yet not: surprising because Gewirth himself has not consciously or overtly recognised, as far as I am aware, the pronounced and crucial Stoic character of his ethical and eudaimonic theory.[6] Although frequently making references to the Aristotelian and Kantian elements in his own theory,[7] he rarely if ever makes overt or even indirect references to similarities or differences between his theory and that of the Stoics.[8]

Before proceeding to examine the close relationship between the Gewirthian neo-Stoic theory I am ascribing to Gewirth, and the traditional ancient philosophy of Stoicism, I will first provide a brief historical summary of some of the central characteristics of Stoic philosophy. A further closer and systematic analysis of those Stoic principles would be provided through my comparative analysis of Stoic philosophy and the parallel Gewirthian neo-Stoic theory that follows.

STOIC PHILOSOPHY

Most of what is known of Zeno's philosophy, the founder of Stoic philosophy, is through the writings of later Stoics, particularly the Roman Stoics, Epictetus, Seneca and the Emperor Marcus Aurelius. Zeno, like all the Stoics, preached equality of the sexes, and also claimed that men and women should dress alike. Observing how unisex fashion has gained ascendancy, one could say that Zeno and his Stoics were in truth the early trend setters.

According to the Stoics, philosophy is like an animal, logic corresponding to the bones and sinews, ethics to the fleshy parts, physics to the soul. Though Stoic writings, especially of the early Stoics, including Zeno, Cleanthes and Chrysippus, covered all topics including logic, theory of knowledge (Epistemology), metaphysics and ethics, it is the latter that has had the biggest and more profound surviving influence, mainly through the works of the later Hellenic and Roman Stoics, notably those of Epictetus, Seneca and Marcus Aurelius. Given also, its contemporary importance and relevance, it is on ethics that the following exposition of Stoic philosophy will primarily focus.

The Life of Virtue and the Life of Wisdom

According to the Stoics, the life of virtue is both necessary and sufficient for living a happy life. Essential to understanding this, is the distinction the Stoics made between those things that are completely within our power and control and those things that are not always and not completely within our control. According to Epictetus (2008, 221),

> Our opinions are up to us, and our impulses, desires, aversions – in short, whatever is our doing. Our bodies are not up to us, nor our possessions, our reputations, or our public offices, or, that is, whatever in not our doing.... So remember, if you think that things naturally enslaved are free or that things not your own are your own, you will be thwarted, miserable, and upset, and will blame both gods and men. (*Enchiridion*, ch.1:1–3)

The only thing over which we have control, according to the Stoics, is our faculty of judgement. Since anything else, including all external affairs and circumstances are not within our power, we should adopt towards them an attitude of indifference. According to Epictetus (2008, 223),

> What upsets people is not things themselves but their judgments about things. For example, death is nothing dreadful (or else it would have appeared so to Socrates), but instead the judgement about death is that it is dreadful, that is what is dreadful. (*Enchiridion*, ch.5)

To avoid unhappiness, frustration and disappointment, we therefore need to do two things: control those things that are within our control (our beliefs, judgements, desires and attitudes) and be indifferent to those things which are not in our control (things external to us). As rational beings, we should therefore perfect our characters through living a virtuous life because it is the only thing that can bring us both *ataraxia* (tranquillity), *autarkia*, (self-sufficiency) and *eudaimonia* (happiness); a happiness that is totally within our control, just because it depends on our own judgement which is entirely within our control.

Associated with the idea of a virtuous life is the idea of wisdom. The ultimate object of philosophy according to the Stoics is to teach us not knowledge, but wisdom, understood as a way of living a virtuous life. Wisdom, therefore, understood in the Stoic sense was a way of life that brought peace of mind (ataraxia), inner freedom (autarkia), happiness, (eudaimonia) and a cosmic consciousness. By "cosmic consciousness," the Stoics understood that quality, as universal reason that permeates the whole cosmos and by virtue of which we all are, as human beings, integrated parts of the cosmos. As Pierre Hadot tells us (1995, 273),

> The exercise of wisdom entails a cosmic dimension. Whereas the average person has lost touch with the world, and does not see the world qua world, but rather treats the world as a means of satisfying his desires, the [stoic] sage never ceases to have the whole constantly present to mind. He thinks and acts within a cosmic perspective. He has a feeling of belonging to a whole, which goes beyond the limits of his individuality.

Most importantly, the Stoics viewed philosophy as therapeutic, intended to cure mankind's anguish. For according to an Epicurean saying, "vain is the word of the philosopher which does not heal any suffering of man" credited to an inscription erected by Diogenes of Oinoanda, province of Lycia, in 120 AD to honour Epicurus.

Social and Political Engagement

Closely associated with the cosmic perspective in Stoic philosophy, is the communal perspective. That is, the concern for living in the service of human community, and for acting in accordance with social justice. Thus, according to the Stoics, philosophy entails a community engagement. This is something that the Stoics have in common with Aristotle who saw philosophy as essentially political. That is, concerned with the affairs of the polis. For the Stoics, however, the polis was the cosmopolis, not the city-state but the whole world.

Significantly, what is missing in most people's lives according to the Stoics is the practice of wisdom understood not as the acquisition of mere knowledge or information but as a way of being in the world that requires one to live a life in accordance with virtue. More than information, wisdom requires transformation. More than knowledge, wisdom requires practical ethics. Some knowledge about the world, even abstract and theoretical knowledge is necessary for wisdom, but it is not sufficient. To become wise, one requires not only to be informed but more importantly to become transformed through the practice of philosophy, specifically Stoic philosophy, as a way of life. This is by no means an easy task and it may in fact prove very difficult for most people. However, the threat to the natural environment through continuous commercial exploitation to satisfy our insatiable desire for consumer goods as well as the constant outbreak of wars around the globe, not to mention, the current and ongoing pandemic across the world, in 2020–2021, renders the ethical lifestyle recommended by the Stoics, with its emphasis on a simple, peaceful, and a frugal but happy way of living, that respects not only the ethical rights of all individuals as citizens of the world but also extends that respect to the whole of the natural environment, as well as the digital environment, as we shall examine later in chapter 4, as ethically desirable.

Philosophy as a Way of Life

The Stoics viewed philosophy not just as an interesting intellectual exercise, but as a way of life available to everyone – the Art of Life or "techne viou." The following is an outline of the reasons of why the Stoics viewed philosophy in this way:

1. The goal in life is to be happy, for who would not want to be happy.
2. Only the possession of what is good can make us truly happy.
3. Moreover, only the good that is always within our control can make us happy.
4. The only good is virtue and moreover virtue is always within our control. For being virtuous, is always up to us. The virtues, among others, sometimes referred to as the cardinal virtues, are justice, temperance or moderation, prudence, or practical wisdom, and courage.

5. Therefore, only the possession of virtue can truly make us happy.
6. The possession of other things like health, wealth, fame, beauty, status, popularity are desirable but are not always good. Their possession can sometimes bring us unhappiness, not happiness. Moreover, their possession can sometimes result in unethical and bad conduct that harms others and causes their unhappiness.
7. Hence, although in normal circumstances things like wealth and status should be preferred, their possession is unnecessary for happiness. They are merely preferred indifferents.
8. Therefore, a virtuous person can be happy without them since only virtue, as the only good, is what one requires to be happy.
9. But why is virtue the only good?
10. Virtue is the only good because only virtue is unconditionally good. For in all circumstances, virtuous conduct is in agreement with nature, and it alone determines what is good.
11. Hence, virtue is the only good because it allows one to live in accordance with nature.
12. But what does to live in agreement with nature, mean?
13. Living in agreement with nature means simply, living a life in accordance with reason for nature is perfectly rational.
14. Therefore, a virtuous life is a life lived in accordance with the perfection of one's reason.
15. And moreover, reason is something that we share with all of nature and God, since God and nature are, according to the Stoics, one and the same, since the Stoics identified God with universal reason that permeates the whole of nature.
16. Thus, in living rationally and virtuously we share in God's perfection.
17. And insofar as philosophy helps us to live a life in which we share in God's perfection, then philosophy is truly the art of life.
18. Philosophy can be viewed as the art of living in two ways: (a) as a craft or art which once learned can enable one to live a good and fulfilling life on a daily basis and (b) the perfection of one's human nature – becoming the best of one's kind – humankind. For the Stoics, the perfection of one's human nature meant the perfection of one's rational nature. For according to the Stoics, the essence of human nature and nature generally, the whole universe in fact, was reason. Thus, in perfecting one's rational nature one becomes at once the object and the subject of art, that is, the art of life. This was the highest form of art, for in perfecting one's rational nature through the practice of philosophy one became like God since God, as we saw, was identified by the Stoics with nature which they viewed as supremely rational and orderly.

In conclusion of this section, philosophy can, according to the Stoics, be conceived as the art of living in these two ways: as a process, an acquired and learned craft whose application can enable one to live happily regardless of the slings and arrows of outrageous fortune; and as a product or an end, namely, the perfection of one's rational nature which we share with God. In this second idealistic and aspirational sense, one can become through the perfection of one's rational character, godlike.

As a point of interest, the story of creation in the Book of Genesis (King James Bible, chapter 3:22), viewed here as an allegory, tells of God expelling Adam and Eve from Eden, as a precautionary measure "lest they eat from the Tree of Life and become like God immortal." The suggestion here in the form of an allegory is perhaps humanity's potential in becoming through wisdom in the form of existential meta-knowledge of how to live one's life, immortal and godlike. We shall return to that suggestion in the final chapter of this book, chapter 9, by exploring the possibility of a beneficial symbiotic relationship, between humans and artificial general intelligent, AI agents. The Stoic idea is that humans, through virtue and living in accordance with nature perceived by them to be divine, can achieve perfect eudaimonia and become part of God. Though conceived differently, the Christian idea of salvation and redemption of humankind after the fall, through Jesus Christ's divine sacrifice, and the Stoic idea of becoming part of God or nature by the perfection of oneself through virtue, though different in content and context, appear to run parallel, around the central idea of humanity's potential for becoming godlike and sharing God's kingdom: in the Bible story, God's Heaven, and in the Stoic story, this Natural World, in which God, whom the Stoics identified with nature, is immanent, ever-present, omnipotent and benevolent. The characteristics traditionally applied by theist philosophers to the Christian God.

COMPARISON OF STOIC PHILOSOPHY AND GEWIRTH'S RATIONALIST PHILOSOPHY

First, let me offer some brief explanatory remarks about methodology. To demonstrate the essential Stoic features of Gewirth's ethical theory, I will juxtapose the main central doctrines of Stoic philosophy with similar parallel precepts and principles present in Gewirth's work. I will refer extensively to relevant passages in Gewirth's book, *Self-fulfillment* (1998), which bears close similarities to Stoic ethics, especially as regards the Stoic views on eudaimonia or happiness, virtue and reason. I will claim and support this claim with textual analysis of relevant key passages, that the whole tenor of *Self-fulfillment* is paradigmatically Stoic in most if not all of its main features. As we shall see, Gewirth's notion of "self-fulfillment" is a neo-Stoic version of Stoic "eudaimonia." By the end of the comparative analysis of the central Stoic ethical doctrines and Gewirth's ethical principles and precepts, we will reach the conclusion, on the basis of that analysis, that Gewirth's ethical theory is essentially neo-Stoic. For it shares most if not all the central doctrines of Stoic ethics.[9] I will say more about that later.

A further methodological point is that my references to Stoic philosophy will primarily rely on secondary sources. Mainly, the work of some leading contemporary Stoic scholars such as, A.A. Long, Gisela Striker, Julia Annas, Martha Nussbaum, Pierre Hadot and last but not least, the work of Laurence Becker, especially with regards to neo-Stoicism. I will refer to extracts from original ancient texts as and when required. I believe that there is sufficient evidence of a general consensus among contemporary Stoic scholars of what the central doctrines of Stoic ethics are. I will thus rely primarily on the work of the aforementioned contemporary Stoic scholars in outlining the central Stoic doctrines and comparing them to similar principles and precepts in Gewirth's work, but will also refer to original sources for further emphasis and explication of key points, when required.

Finally, a related issue to the one above is whether there was a consensus among the ancient Stoic philosophers themselves as to what constituted the central doctrines of Stoicism, specifically Stoic ethics. Were the central doctrines of Stoicism common among the early Stoics such as Zeno of Citium (342–270 BC) and Chrysippus (280–207 BC) of the early stoa, the Stoics of the middle stoa such as Panaetius (180–109 BC) and Posidonius (135–151 BC), and the much later Stoics of the Roman Empire such as Seneca (4BC–AD 65), Epictetus (55–135 AD) and Marcus Aurelius (121–180 AD). There were indeed some differences between all these Stoics, with the later Stoics considered to have adopted a more moderate and less stringent ethical doctrine than the early Stoics such as Chrysippus. However, as Gisela Striker has pointed out, notwithstanding those differences, there appears to have been a common core of doctrines among all those Stoic philosophers, corresponding more closely perhaps to the doctrines developed by Chrysippus, generally considered the most important Stoic philosopher of antiquity after Zeno of Citium.[10] For the purpose of this chapter, I will follow Striker on this issue and adopt the view that there is in fact a common core of doctrines so fundamental to Stoicism that they can be attributed to Stoic philosophy as a whole. I will only highlight differences among particular Stoic philosophers when this is of significant relevance to my comparative analysis of the Stoics and Gewirth.

The central doctrines and other features of Stoic philosophy selected for discussion and comparison to Gewirthian ethics, are the following: (1) "the final goal"; (2) "a life in agreement with nature"; (3) *"eudaimonia"*; (4) "virtue"; and (5) *"oikeiosis."* As mentioned earlier, my selection of the central doctrines of Stoic philosophy is based primarily on a number of representative secondary sources[11] from contemporary Stoic scholarship that support a consensus on what constitutes the central doctrines of Stoic ethics. This consensus concurs with my own general reading of Stoic texts and Stoic philosophy.[12] In this regard, I consider that selection to be commonplace and uncontroversial.

For easy comparison, the central doctrines of Stoic ethics have been juxtaposed with the parallel and corresponding central features and precepts of Gewirth's rationalist philosophy and in particular ethics. The comparison is intended to demonstrate the close similarity between Stoic and Gewirthian ethics thus establishing my claim that Gewirth's rationalist ethical theory is essentially and paradigmatically neo-Stoic.

The Final Goal

A. A. Long[13]

The Stoics[14]

The Hellenistic philosophers, including the Stoics, claimed that *eudaimonia* or happiness was the ultimate goal of a good life, though each of the Hellenistic schools identified happiness with something different: the Epicureans identified it with pleasure and the avoidance of pain, the Pyrrhonian Sceptics with suspension of all intellectual judgement as the only way to achieve *ataraxia* or tranquility, and the Stoics exclusively with a life of virtue, which they, in turn, identified with a life in agreement with "nature." The Stoics identified nature with the whole cosmos or universe, which they viewed as a perfectly ordered and rational deterministic system; a system to which we are intrinsically related as parts, by virtue of being inherently rational beings. A life in accordance with nature was thus a life in accordance with reason, one which human beings as inherently rational shared with the whole of nature as related parts. It is, so to speak, our "nature" to be rational, and it is our rational nature which leads us, through various intermediary stages, to a virtuous and ethical life that ultimately results in *eudaimonia* or happiness.[15] For a virtuous life, as the only good, was considered by the Stoics to be both necessary and sufficient for a good and happy life. Becoming truly virtuous, however, required the perfection of one's rationality and character over a lifetime, and that was an achievement *(katorthoma)* reserved for those capable of becoming Stoic sages. To this extent, Stoic ethics is idealistic, the sage is perceived as the ideal limiting case that other aspiring Stoics can gradually approach but only rarely reach. For the Stoics, Socrates was the prime exemplar of a proto-Stoic. Rational and moral perfection that ultimately leads to eudaimonia is therefore, in a sense, utopian, but nevertheless, worth pursuing.

The above is an attenuated outline of some of the central elements of Stoic ethics. Each of those elements will be discussed further in more detail in the following sections. A subtle point that can sometimes be overlooked with regard to *eudaimonia* viewed as the final goal, is that for the Stoics it was strictly speaking *virtue* not happiness that was considered the goal. That is, the target of the good life for a Stoic was not in the first instance to become happy as such, but to become perfectly virtuous; for virtue was the only good worth pursuing and ultimately worth having. However, as a necessary and sufficient prerequisite for happiness, a Stoic who achieved perfect virtue will also be perfectly *eudaimon,* or happy. Thus, in a sense *eudaimonia* or happiness was the outcome of becoming and being perfectly virtuous, not the immediate target for a Stoic. For by being virtuous one essentially lacked nothing, as virtue was the only good worth having. And if one lacked nothing one would be completely self-fulfilled and happy. Happiness was the trophy for leading a virtuous life. Virtue becomes precisely its own reward, for its necessity and sufficiency for happiness logically entails and causally guarantees happiness for the person who possesses it.

Gewirth[16]

Towards the beginning of *Self-fulfillment,* Gewirth poses (and provisionally answers) the following question: "Is there any general object of all aspirations? Happiness might seem to be a plausible answer" (1998, 46). In anticipating his general thesis in his book, namely that a moral and virtuous life is essential for self-fulfillment (the question whether for Gewirth a moral and virtuous life is also sufficient for self-fulfillment will be discussed separately below), Gewirth claims that the "basic question for self-fulfillment as capacity-fulfillment[17] is: How can I make the best of myself? Put affirmatively, the general formula of such fulfillment may be stated as follows: For you to fulfill yourself is for you to achieve the best that it is in you to become" (1998,59).

And for Gewirth, "achieving the best that it is in you to become" – self-fulfillment considered as *capacity fulfillment* – "*is* indissolubly bound up with ... morality" (1998, Preface). Gewirth's moral conception of self-fulfillment as the process of becoming the best one can, *qua* human being, is clearly very close to the Stoic ideal of moral perfection as an essential prerequisite to happiness or self-fulfillment. I believe there is ample textual evidence in Gewirth's *Self-fulfillment* to draw a close parallel between Stoic *eudaimonia* and Gewirthian self-fulfillment. This conceptual equivalence will become more evident as the argument advances and develops. For Gewirth, the final goal in life is self-fulfillment as capacity fulfilment. According to him, self-fulfillment [is] construed as a final end of aspirations and capacities. Even if the attainment of self-fulfillment is best assured when it is a "by-product" of other goals to which one is directly committed . . . it is still valued for itself, because it consists in carrying to fruition, by a self-knowing, self-reflective process, one's worthiest capacities (1998, 223, and 46, 59, 182–89, and 199).

From our discussion so far, it is clear that Gewirth's notion of *self-fulfillment* as capacity fulfilment, has all the essential characteristics of Stoic *eudaimonia*. What is not yet clear is whether virtue is not only necessary but also sufficient for self-fulfillment, as is it for Stoic *eudaimonia*. I will defer the discussion of this issue for later. In the meantime, setting aside the *sufficiency thesis* for happiness, I will regard Gewirth's notion of self-fulfillment as capacity-fulfillment equivalent to Stoic *eudaimonia*.

Moreover, I will use the term "happiness" interchangeably for both Gewirth's "self-fulfillment" and the Stoic notion of *eudaimonia*.

A Good Life in Agreement with Nature

The Stoics[18]

> As Cicero explains, the value of anything in Stoicism is defined by reference to Nature.(Long 1986, 198)

> So, the Stoics are telling us that the goal of life, happiness, consists in a conscious observation and following of nature's will. (Striker 1996, 224)

According to the Stoics, happiness consisted in living a virtuous life and a virtuous life was one in agreement with nature, which they identified with divine reason – which, permeates the whole cosmos or universe. Since virtue was the only good, a good life was a life in agreement with nature. And since nature was identified by the Stoics with reason, they identified a life in agreement with nature with the perfection of human reason. Moreover, it is only by living a perfectly rational and virtuous life (what the Stoics considered, a *good life*) that human beings can truly become happy. Hence, for the Stoics there is an intrinsic and integral relation between a virtuous life, rational agency, nature, and happiness or self-fulfillment, all of which relate to a good life.

Gewirth[19]

As we saw from my exposition of Gewirth's model of self-fulfillment, self-fulfillment as capacity fulfillment consists in developing one's capacities and making the best of oneself. In attaining this, it is essential to act in accordance with human rights as required by universal morality on the basis of the principle of generic consistency (PGC). For *reason*, which Gewirth identifies as the best of all human capacities, requires that we act in accordance with the requirements of universal morality. Moreover, it is reason's "veridical capacity" (Gewirth 1998, 72) that reveals to rational agents that it is the natural property of *purposiveness* that they all have in common that provides the natural foundation for human rights and universal morality, and it is that which underlies the ascription of inherent dignity to human agents (Gewirth 1998, 222). Furthermore, reason also reveals to us that the development and inculcation of the virtues is both an important means and an essential constituent for both a good and a self-fulfilled life (Gewirth 1998, 115, 135). In Stoic ethics, as in Gewirthian ethics, there appears to be an intrinsic and integral relation between the following components that together constitute a good life: a virtuous life, rational agency, "nature" understood in terms of the natural property of purposiveness, and happiness or self-fulfillment. As Gewirth tells us, "the components of self-fulfillment . . . have all derived in various ways from the root idea of the self as a rational purposive agent" (Gewirth 1998, 217).

In Gewirthian terms, a life in agreement with nature is a life in agreement with the perfection of rational purposive agency, which, in turn, is, or ought to be, in agreement with universal morality, which is essential for a good and self-fulfilled life. In contrast to the Stoic conception of universal nature as divine, however, Gewirth's notion of "nature" is *human* nature, understood as rational purposive agency. But for the Stoics (Long 1986, 174) as for Gewirth, the concept of human nature is both descriptive and prescriptive. Interestingly, A. A. Long tells us that for Panaetius, one of the later Stoics of the middle-stoa, human nature rather than the universal nature of the earlier Stoics was the primary interest (Long 1986 211). Strictly speaking, in this respect, there appears to be a closer link between Gewirth and Panaetius than between Gewirth and the earlier Stoics such as Chrysippus and later the Roman Stoics, such as Epictetus and Marcus Aurelius. Gewirth's notion of "nature" as rational purposive agency is, in that respect, more appealing to our contemporary naturalistic temperament because it avoids the methodological problems that the Stoic metaphysical view of nature poses. Hence, Gewirth's neo-Stoicism, which I have formulated and have ascribed to him (Spence 2016, Chapter 10) at least in the form I am attributing to him, has a significant methodological advantage over Stoicism. One could say that Panaetius, with his more naturalistic view of nature, was anticipating Gewirth's neo-Stoicism.

Eudaimonia

Happiness . . . is the polestar of our ethical theory (Becker 1998, 18).

The Stoics[20]

According to the Stoics, eudaimonia or happiness is not something that happens to you by chance, like winning the lottery, but rather something that one *achieves* through becoming perfectly rational and virtuous. Only few people can ever achieve eudaimonia because only a few people can become truly virtuous. At the limit, the sage is viewed by the Stoics as an ideal exemplar; someone to be looked upon as a role model for those aspiring to become happy through becoming wise and virtuous. Becoming happy or eudaimon for the Stoics was thus considered a *katorthoma* or supreme achievement. Something earned, not something given. But though difficult, the attainment of happiness through virtue was not only practically possible but also prescribed by nature; in particular, rational human nature. Insofar as becoming virtuous was within everyone's control, so was happiness.

Gewirth[21]

In *Self-fulfillment*, Gewirth distinguishes between self-fulfillment as *aspiration fulfillment* and self-fulfillment as *capacity fulfillment*. The former he identifies as the satisfaction of one's deepest desires, the latter as the process and goal of making the best of oneself. The difference between the two as Gewirth claims,

> is that aspirations and their fulfillment are tied more closely to person's actual desires, while capacity-fulfillment bears more on making the best of oneself and thus serves as a normative guide to what desires one ought to have, where this 'ought' may (but need not) go beyond person's actual desires (Gewirth1998, 107).

Gewirth goes on to identify these two distinct conceptions of self-fulfillment with two other conceptions of happiness thus, identifying "self-fulfillment" with "happiness." He goes on to say that the putative connection of his two notions of self-fulfillment as aspiration fulfilment and capacity fulfilment with happiness "brings out further why self-fulfillment is so highly valued as a superlative condition of the self" (Gewirth1998, 14–15). Clearly, it is Gewirth's notion of capacity-self-fulfillment that is equivalent to Stoic eudaimonia.

From a close examination of *Self-fulfillment*, it becomes clear that Gewirth's notion of self-fulfillment as capacity fulfilment has all the essential characteristics of Stoic eudaimonia. What is not yet clear is whether virtue is, as is it for Stoic eudaimonia, not only necessary but also sufficient for self-fulfillment. I will defer discussion of this issue for later. In the meantime, setting aside the sufficiency thesis for happiness, I will regard Gewirth's notion of self-fulfillment as capacity fulfilment equivalent to Stoic eudaimonia. In claiming this, I take Gewirth's notion of self-fulfillment as capacity fulfilment to be, with regard to its equivalence to Stoic eudaimonia, a *second-order self-fulfillment*. That is, self-fulfillment, qua rational human being.

A person can attain *first-order self-fulfillment* by becoming the best he can possibly become, given his particular aspirations and capacities, in a certain specific activity, for example, the best tennis player. However, that person may nevertheless fail to become self-fulfilled at the level of second-order self-fulfillment because he does not succeed in becoming the best person, qua rational human being, he can possibly become. The attainment of such second-order self-fulfillment, which is equivalent as I claim to Stoic eudaimonia, would only be possible for a Gewirthian neo-Stoic sage and his Stoic counterpart. Lesser mortals who do not succeed in becoming perfectly rational and virtuous in all their life plans and activities, would be able to attain *first-order self-fulfillment* in their specific activities, but would only be able to attain *second-order self-fulfillment* by first becoming the best of oneself, or in my analysis of Gewirth's capacity-self-fulfillment as equivalent to Stoic eudaimonia, becoming a neo-Stoic sage.

For the purpose of this chapter, I will use the term "happiness" interchangeably with both Gewirth's notion of "self-fulfillment" and the Stoic notion of "eudaimonia." Self-fulfillment as capacity fulfilment is the primary focus of Gewirth in *Self-fulfillment*. As for the Stoics, so too for Gewirth, self-fulfillment as capacity fulfilment is an achievement, one that requires, as in Stoic ethics, a life lived in agreement with both reason and morality. For in both Stoic and Gewirthian ethics, human rational nature as an integral part of a rational and divine universe (as in the case of the Stoics), or as rational purposive agency (as in the case of Gewirth), prescribes that one should make the best of oneself. And this, in turn, requires, at a minimum, living both an ethical and a virtuous life. According to Gewirth, "personalist morality . . . gives counsel and precepts for the self's having a good life through personal development of one's capacities whereby one makes the best of oneself" (Gewirth1998, 107).

It is in self-fulfillment, through its relation to purposiveness and the fulfilment of capacities in making the best of oneself, that Gewirth locates the meaning of life. It is not, as in Stoicism, a metaphysical meaning that emanates from a divine and rational all purposeful universe, but rather a human all too human meaning, one that emanates from our own rational purposive nature. It is this inherent purposiveness aligned to our rationality that drives us on to the perfection of our individual selves and not some promise of a metaphysical reward in this life, as in the Stoics, or a reward in some other transcendent life as in Plato. Gewirth's self-fulfillment is wholly naturalistic. For Gewirth, "the meaning of life . . . consists in the pursuit and attainment of the values of personalist morality as guided by the rational justification of universalist morality and the analysis of freedom and well-being as central to the highest development of the virtues based upon these necessary goods of action"(Gewirth1998, 189).

However, for practical purposes, both Stoic eudaimonia and Gewirthian self-fulfillment amount to the same thing: a quest for making the best of oneself in agreement with one's intrinsic rational human nature that accords with an ethical and virtuous life – one that ultimately leads to eudaimonia or self-fulfillment. As in Stoicism, so to in Gewirth's ethical schema, self-fulfillment is a "maximalist" concept (Gewirth1998, 216), for it involves the perfection of oneself through the perfection of one's human rational nature.[22] And as in Stoicism, so too in Gewirth's ethical theory, self-fulfillment is something to be aimed at as an ideal, a practically possible target, but one that may never be completely reached. As Gewirth says, "self-fulfillment is far more a process than a finished product" (Gewirth1998, 226).

It seems that for Gewirth as for the Stoics before him, self-fulfillment is something that perhaps only a sage can achieve. For mere mortals like the rest of us, the process is all one can hope for. As the Greek poet Constantine Cavafy remarked in a highly suggestive poetic line (*Ithaca*), it is the journey itself that ultimately counts. For in undertaking it, we stand to become wiser if not completely wise. This perhaps accords with what Gewirth tells us, "there is no climactic nirvana, but there can be sequences of self-improvement that overcome the effects of alienation and achieve cherished values. One's best is never finalized, but it can be more fully approached" (Gewirth1998, 227).

Virtue[23] and Eudaimonia[24]

The Stoics

The Stoics Definition of Virtue

The following is a definition of virtue from A. A. Long (1986, 92).

(i) "Virtue is not defined by the consequences in the world which it succeeds in promoting, but by a pattern of behavior that follows necessarily from a disposition perfectly in tune with nature's rationality.
(ii) The right thing to do is that which accords with virtue, and this is equivalent to saying that it accords with the nature of a perfect rational being.
(iii) Virtue accords with nature in the sense that it is the special function or goal of a rational being to be virtuous."

The Stoic Relationship of Happiness to Virtue

Another central feature of the Stoic conception of happiness is its intrinsic relation to virtue. For as noted earlier, virtue as the primary comprehensive goal of human nature is not only the essential means for the attainment of happiness, but more importantly, it is constitutive of it (Long 1986, 197). In being the end or goal of life, virtue is both the means and the end of a happy life. The means, for the development and inculcation of virtue leads to happiness; and the end, because its possession as the only good guarantees a good life; for virtue is constitutive of happiness and its possession as the only good necessarily guarantees one's happiness. Another related characteristic of happiness is its self-sufficiency; for insofar as virtue as the only good is entirely within one's control, and that alone is sufficient for happiness, then happiness results from virtuous self-sufficiency (Long 1986, 234).

An important qualification, however, must be made to the view that happiness and virtue in Stoic ethics are so intimately and intrinsically connected, that one cannot tell them apart, and as a result, one might then be led to view Stoic happiness as nothing more than a life of virtue. This would be a mistake, however. For as Becker correctly points out, Stoic virtue is not the same as say Kantian duty (Becker 1998, 150–51). There is joy in Stoic ethics over and above the state of just being virtuous and always doing the right thing out of a grudging sense of duty. In Stoicism, the possession of virtue fills one's soul with joy similarly to how a state of health and fitness fills one with a general sense of well-being, both physically and psychologically. The Stoic sage, as the sage in Plato's *Symposium* standing at the top of the ladder of human perfection and gazing at the form of the good, is in love with the good[25] and that alone fills him with abundant joy.

Julia Annas (1993, 329–35), rightly insists, contrary to the view that Stoic eudaimonia is nothing more than a life of virtue, that we must take Stoic eudaimonists at their word for ultimately what they are offering us is an account of a happy life and not just a virtuous one. And these two things are, though intrinsically connected, conceptually distinct, as are shape and size, for example, although shape and size are intrinsically connected, they are nevertheless, conceptually distinct. Insofar as virtue is the only good and virtue is entirely within one's control, then possession of the only good is both necessary and sufficient for happiness and potentially, within the reach of everyone. Moreover, sages like Socrates and the Buddha are real exemplars that render this view not a mere philosophical abstraction but a real practical possibility.

Stoic Virtue as both Necessary and Sufficient for Happiness

From the two propositions that human virtue is rational perfection, and that the perfection of one's reason is in agreement with nature, the Stoics inferred that virtue is good, the only good, for it alone allows one to live in agreement with nature (Striker 1992, 1209). Moreover, being a function of human rationality virtue is within one's rational control and as the only good, its possession leads to happiness or eudaimonia. That virtue is the only good follows from the view that only virtue is unconditionally good. For in all circumstances, virtuous conduct in agreement with nature, alone determines what is good. Hence, virtue is the only good because it allows one to live in accordance with nature.

Whereas health, wealth, friends and social status might be good, at least some, or most of the time, what the Stoics referred to as "preferred indifferents" (the Stoic term for contingent "goods"), they are not always good as their possession may not result in a virtuous life in agreement with nature. Worse still, they may lead one to unethical and vicious actions that are bad because such actions are not in agreement with nature; on the contrary, they are in disagreement with nature because, ultimately, they are inconsistent with one's rational nature.

The above analysis is, I believe, in keeping with the general features of Stoic ethics concerning the sufficiency of virtue for happiness.[26] What emerges is the following schema:

1. Virtue is the perfection of one's rational nature.
2. Virtue as the perfection of one's rational nature allows one to live in agreement with nature.
3. As the only thing that allows one to live a life in agreement with nature, virtue is the only good.
4. As such, virtue is unconditionally good.
5. Only possession of what is unconditionally is good can render one happy.
6. Therefore, since only virtue is unconditionally good, only virtue can make one happy.
7. Virtue is completely within every rational person's control.
8. Hence, every rational person can potentially (because everyone is by nature inherently rational) become happy and, moreover, must aim at becoming happy by becoming virtuous.

The above eight-point outline renders, I hope clearly, the intrinsic relationship between the central features of Stoic ethics: reason, virtue, nature and eudaimonia. Moreover, the outline draws attention to both the descriptive as well as the prescriptive characteristics of Stoicism. Furthermore, it highlights the ideal and developmental aspects of Stoic ethics. However ideal, the status of sagehood, as the only possible state for true and lasting happiness, remains a practical possibility for every rational person. In the *Tusculan Disputations*, Cicero devotes the whole book (V) to a defence of the necessity and sufficiency of virtue for happiness. Laurence C. Becker has identified the following premises as the key premises in Cicero's argument (Becker 1998, 152):

(1) "No one can be happy except when good is secure and certain and lasting (V.xiv.40).
(2) The good of virtue is secure and certain and lasting because (a) once achieved, its maintenance is within the agent's control (V.xiv.42), and (b) it is free from the disturbances of the soul that produce wretchedness (V.xv.43).
(3) Moreover, in its affective dimension, a virtuous life is characterized by tranquility and joy, and thus may unproblematically be described as a happy life (V.xv.43)
(4) No form of happiness can be good unless it includes, or is founded upon, virtue, or what is right" (V.xv.44–5).

Gewirth: *Definition of Virtue*

It will be recalled that a Stoic definition of virtue is that "it is not defined by the consequences in the world which it succeeds in promoting but by a pattern of behavior that follows necessarily form a disposition perfectly in tune with Nature's rationality" (Long 1986, 192). This definition of Stoic virtue is very much in keeping with Gewirth's own understanding of virtue: namely, a "disposition perfectly in tune with human rationality" where *human nature* as rational purposive agency replaces the metaphysical notion of Stoic *Nature*. And for Gewirth, so too for the Stoics, "the right thing to do is that which accords with virtue, and this is equivalent to saying that it accords with the *nature of a perfect rational being* (emphasis added)" (Long, 1986, 192). For Gewirth, the right thing to do accords with the requirements of the PGC, which, in turn, accords with the natural property of "rational purposive agency." That, in turn, requires the development and inculcation of the virtues as the enabling dispositions for acting and living ethically. And acting and living virtuously then accords, at least for the neo-Stoic sage, with the nature of a rational purposive agent. And being a common natural property that all rational agents have in common, rational purposive agency is therefore universal.

Gewirth's Relationship of Happiness to Virtue

Gewirth's attention to the virtues and their relationship to both rationality and morality, mediated through the PGC, is extensive and comprehensive. What will concern us in this section is Gewirth's account of the relationship between virtue and self-fulfillment. The relationship as expressed by Gewirth appears to be as follows: insofar as morality is intrinsically related to self-fulfillment as capacity fulfilment (Gewirth 1998, 78–79), and insofar as the virtues are essential in enabling rational agents to act and live ethically in compliance with the requirements of the PGC (Gewirth 1998, 135), then the virtues are also intrinsically related to self-fulfillment (Gewirth, 1998, 87). The above-described relationship between virtue and self-fulfillment appears, in the first instance, to be instrumental. That is, virtues are essential because they are effective means for ethical development thus enabling one to comply more effectively with both the negative and positive requirements of the PGC.

This instrumental understanding of the virtues is also part of the developmental aspect of Stoic ethics. By aiming and attempting to act and live virtuously, one gradually becomes inculcated in virtuous conduct through habituation or *oikeiosis*. The Stoic concept of *oikeiosis*, will be discussed in more detail below.[27]

However, there is also, as in Stoic ethics, a constitutive aspect of the virtues in Gewirth's model of self-fulfillment. Once one becomes habituated to acting ethically in compliance with the PGC, then virtuous conduct becomes, as it were, one's *second* nature. In the Gewirthian neo-Stoic sage, at the limit of virtuous perfection, however, virtue becomes, together with rationality, the sage's *first* and only nature. The perfection of reason and virtue combine and match in the neo-Stoic sage, and though remaining conceptually distinct, become in the sage's actions, *practically* indistinguishable. Thus, the neo-Stoic sage acts virtuously not only because his reason demands it, but more importantly because his reason is attuned to virtuous conduct. Unlike the Kantian "sage," the Gewirthian neo-Stoic sage, like his Stoic counterpart, does the right thing out of love for the good and not out of a joyless sense of duty. The neo-Stoic sage, like his Stoic predecessor, enjoys being virtuous and this, unlike the Kantian sage, does not in any way diminish his goodness or his happiness.

The *developmental* and the *constitutive* aspects of virtue are illustrated by Gewirth in the following passage: "There is a difference, however, between one's having a duty to act form a virtue and one's having a duty to try to develop a virtue. It is the latter that is required for capacity-fulfillment" (Gewirth 1998, 135). When, however, one becomes self-fulfilled, by becoming perfectly virtuous qua a rational person, then one's developmental virtue becomes constitutive both of his character and of his happiness or self-fulfillment. Acting out of virtue for a Gewirthian neo-Stoic sage, as for the Stoic sage of antiquity, is acting out of his perfected rational nature. There is, in contrast to the Kantian "sage," no separation in the Stoic sage between the two distinct thoughts "this is what I *want* to do" and "this is what I *ought* to do." Acting out of desire and acting out of virtue become in the neo-Stoic and Stoic sage alike one and the same thing.

Gewirthian Virtue as both Necessary and Sufficient for Happiness

Insofar as health, wealth, fame and social status, some of the Stoic preferred indifferents, are not essential for complying with the requirements of the

PGC, and moreover, under certain circumstances their possession may result in the violation of those requirements, then virtue would appear to be, as for the Stoics so too for Gewirth, the only unconditional good. Unconditional, because only virtue is under all circumstances, at least in principle, *capable* of enabling a rational purposive agent to act in accordance with the negative and positive requirements of the PGC. I say only *capable*, because like rationality, virtue cannot motivate a sociopath or a psychopath to behave ethically. This is partly because the sociopath's and the psychopath's rational nature has, in some way, become dysfunctional. But even if virtue as the only good, understood as the only unconditional good that is capable of enabling a rational purposive agent to act and live ethically is necessary for self-fulfillment, is it sufficient?

In my book (Spence, 2006, Ch. 10), I explore in detail that question and conclude that as in Stoic philosophy so too in the neo-Stoic philosophy I am attributing to Gewirth in this chapter, virtue is indeed both necessary and sufficient for both neo-Stoic *eudaimonia* and Stoic *eudaimonia*. Constrained by space, I am not able to include in this chapter the full detailed analysis I provide in my book (Spence 2006, 428, 433). However, I will provide in summary an outline of the conclusion of the argument I reach in my book. Namely, that *constitutive* virtue[28] is sufficient for ideal *self-fulfillment* and equivalently *eudaimonia* in Gewirthian neo-Stoic philosophy so long as a minimal rational purposive agency can be preserved and maintained.[29] This will, in turn, require the presence of the material, personal and social conditions that are sufficient for the preservation and maintenance of rational agency, nothing more and nothing less. An interesting contrast that seems to support the social dimension of the neo-Stoic sagehood I have outlined above is a passage from Julia Annas: "virtue is sufficient for non-unhappiness, but for happiness the person must be able to use, not merely possess, virtue" (1999, 45).

Oikeiosis

The Stoics[30]

The transition from developmental to constitutive virtue, from imperfect and incomplete virtue to perfect and complete virtue, requires a kind of ethical transformation. In an insightful passage, Julia Annas describes quite clearly the kind of ethical transformation that Stoic happiness requires:

> As a theory of happiness [the Socratic theory of happiness in Plato], it is very demanding. You won't have got your account of happiness right until you achieve the right perspective, and that will be utterly different from your initial standpoint. Becoming virtuous requires *transforming* (emphasis added) your

values and priorities; nothing short of this will do. All your valuations of conventional goods and evils will turn out to be wrong . . . we are used to moral theories that are demanding, but it is not easy for us to conceive of such a theory which is a theory of *happiness*, rather than something else. But this is not an argument against a theory like Plato's. It shows merely that it takes a long intellectual journey away from common sense and conventional beliefs to achieve the wisdom of the virtuous person and thus to have your whole life, and your conception of happiness transformed. (Annas 1999, 50–51)

Gewirth also recognises and acknowledges the transformative aspect of his concept of ideal self-fulfillment. As he tells us, "to recognize and accept these requirements in one's thought and action may involve a kind of *self-transformation* (emphasis added) whereby one becomes more fully a moral and social being" (Gewirth1998, 86).

It is as difficult for the Stoic and neo-Stoic sage to accept the commonsense view of happiness, primarily the hedonistic view of happiness, as it is for the ordinary person to accept the Stoic and neo-Stoic view of happiness as a view of "happiness." And the difficulty each has in comprehending the attractiveness of each other's notion of happiness lies in their vastly different ethical perspectives. It is for this reason that a practical and gradual transformation must take place over a long period of a person's life before a hedonistic perspective on happiness can be transformed and replaced in the person aspiring to Stoic or neo-Stoic sagehood, with a eudaimonic perspective of happiness.

It is the realisation of the practical necessity for such transformation and the need for providing a conceptual account of how such a transformation *can practically* take place, that the Stoics developed their celebrated theory of *oikeiosis*; a theory which is considered to be one of the most crucial and fundamental doctrines of Stoic ethics. The theory purports to show how a person can progressively develop, because he is naturally conditioned and thus capable of doing so, from the state of having and being motivated by purely hedonistic desires as a child, to the state of acquiring, over a time period of biological, psychological, social, but more importantly rational development, an ethical perspective.

Gisela Striker defines "oikeiosis" as "appropriation" or "familiarization" with oneself and others (Striker 1992,1209), or as "recognition and appreciation of something belonging to one" (Striker 1996, 281). In a similar vein, A.A. Long defines it as the disposition of being "well-disposed towards oneself" (Long 1986, 172). The following passage from Diogenes Laertius is illustrative of the Stoic notion of *oikeiosis* (Long 1986, 185–86):

> The Stoics say that an animal has self-protection as the object of its primary impulse, since Nature from the beginning endears it to itself, as Chrysippus says

in his first book *On goals*: "The first thing which is dear to every animal is its own constitution and awareness of this; for it was not likely that Nature gave it no attitude of estrangement or endearment. It follows then that having constituted the animal Nature endeared it to itself; thus, it is that the animal rejects what is harmful and pursues what is suitable (or akin) to itself."

The assertion that pleasure is the object of animals' primary impulse is proved to be false by the Stoics. For pleasure, they claim, if it really exists, is a secondary product when and only when Nature by itself has searched out and adopted the things which are suitable to the animals' constitution as such pleasure is like the flourishing of animals and the bloom of plants. Nature made no absolute distinction between plants and animals, for Nature directs plants too, independently of impulse and sensation, and in us certain processes of a vegetative kind take place. But since animals have the additional faculty of impulse, through the use of which they go in search of what is suitable to them, it is according to Nature for animals to be directed by impulse. And since reason in accord with a more perfect prescription has been bestowed on rational beings, life according to reason rightly becomes in accordance with their nature; for reason supervenes as the craftsman of impulse.

The following (premises 1–4) is an outline of the conceptual structure of the argument for *oikeiosis* that reflects the crucial features of that concept (Striker 1992, 1209–10):

(1) The first of the two impulses, the impulse for self-preservation, "fosters a self-regarding behavior such as looking after one's health and strength, getting food and shelter, acquiring material wealth, and also developing one's capacities for reasoning and trying to acquire knowledge – ways of acting that one might not ordinarily describe as being morally valuable."

(2) The second impulse, the impulse for sociability, was described by the Stoics as the foundation of justice, "based on respect for others and the desire to help fellow human beings seen as belonging to one, at least in kind."

Following these two material impulses, the impulse for self-preservation and the impulse for sociability or caring for others, "would lead a human being to pursue and avoid things in a systematic, orderly way that can be described as "appropriate conduct."

(3) However, a person who led a "natural life" in this sense would not yet be considered virtuous. For virtue is not only a matter of acting correctly, but more importantly a matter of acting for the right reasons. According to the Stoics, virtue must be based on reason and knowledge, not on instinct. When eventually reason takes control over the natural impulses in a person's life, the person is:

able to recognize that his natural and appropriate conduct exhibits a kind of rational order that is itself far more valuable than the objects pursued in his individual actions. He comes to realize that goodness consists of perfect rationality, and thereupon makes it his ultimate aim to live according to the rational order of nature, of which human conduct is only a minor part. He will therefore cease to value natural advantages in their own right and continue to pursue them only in order to achieve agreement with nature and thus perfect his rationality. *The switch from the pursuit of natural aims to an exclusive interest in rational agreement with nature marks the transition to virtue* (emphasis added).

The "switch" in Gisela Striker's informed passage, marks a similar parallel "switch" in Gewirth's argument for the PGC. Namely, the crucial transition from self-regarding prudential rights to other-regarding moral rights mediated through the recognition that one has those rights on the sole basis of being a rational purposive agent.

(4) Finally, the virtuous person, the Stoic sage, "is the one whose every action is informed by his insight into the will of Zeus [or Nature] and whose only desire is to conform to nature's law."

The four stages described above comprise the central developmental stages in the Stoic doctrine of oikeiosis. Namely, the impulse for *self-preservation*; the impulse for *sociability*; and the recognition that one should not merely act appropriately towards oneself and others on the basis of "natural impulses" alone, but more importantly should do so on the basis of his *rationality*. And finally, the disposition and the practice of always acting according to one's perfected reason and virtue in agreement with nature, results in the attainment of *eudaimonia*. For this is in accordance with one's own rational nature that reflects and is part of, the rational order of nature, or the universe. Acting rationally is appropriate because it is in agreement with nature and agreement with nature is and ought to be the goal of life.

Gewirth[31]

Stages (1)–(3) of the Stoic doctrine of oikeiosis, as outlined above, are parallel to and schematically represent the developmental *virtue* of neo-Stoicism as I described it earlier. Stage (4), the final stage of sagehood, is parallel to and schematically represents the *constitutive virtue* of Gewirthian neo-Stoicism. All four stages of the Stoic doctrine of oikeiosis parallel and schematically represent in Stoic terms, the direct as well as the indirect applications of the argument for the PGC. Stage (4), the stage of sagehood, parallels and schematically represents in Stoic terms, Gewirth's model of ideal self-fulfillment and my ascription to him of the notion of a Gewirthian neo-Stoic sage;

someone who makes the transition from developmental virtue to constitutive virtue and always acts and lives in accordance with perfect rationality and virtue on the basis of both the direct as well as the indirect applications of the PGC.

To be sure, Gewirth's argument for the PGC, both as regards its direct as well as its indirect applications, especially those concerning Gewirth's model of self-fulfillment, is not exactly the "cradle argument" of Stoic oikeiosis, for the earliest stage of childhood development is missing from the structure of the argument for the PGC. However, be this as it may, there is nothing in the formal structure of the PGC that cannot refer, retrospectively as it were, to this earlier stage of childhood development in Stoic oikeiosis. Granted that Gewirth's argument starts with mature rational agents (mature enough insofar as they are capable of minimally reasoning both inductively and deductively), there is no methodological reason why the argument could not, at least notionally, refer retrospectively to that earlier childhood development in Stoic oikeiosis. My comparison of Gewirth's argument for the PGC to the Stoic argument for oikeiosis takes up the comparison at a later stage of Stoic oikeiosis but still following the same pattern and rationale as the argument for oikeiosis.

The following is my conceptual construction in outline of a parallel "Gewirthian Neo-stoic *oikeiosis*" (premises I–III):

(I) Gewirth's argument for the PGC starts, as the Stoic argument of oikeiosis does, at the first stage of human development, namely, the impulse for self-preservation. It will be recalled that in the first stage of Gewirth's argument for the PGC, the agent is only concerned with his own particular purposes. It is his self-regarding concern for his own purposes that initially motivates the instrumentally rational agent to claim self-regarding *prudential* rights to his own freedom and well-being. This first stage of Gewirth's argument for the PGC parallels the first stage in the Stoic argument of oikeiosis. In both arguments, the agent's concern is primarily for himself and for his own desires, interests, and purposes. The claiming of prudential rights by the agent in Gewirth's argument is motivated by self-interest which parallels, at least schematically if not in detail, the self-preservation impulse of the agent in the Stoic argument.

(II) The second and crucial stage in Gewirth's argument for the PGC begins with the agent's realisation that he has prudential rights to his freedom and well-being just by virtue of being a purposive agent. It is this realisation that mediates the crucial "switch" in the agent's internal dialectical perspective from self-regarding prudential rights to other-regarding moral rights.

This stage in Gewirth's argument is parallel to stages (2) and (3) in the Stoic argument for oikeiosis outlined above. Importantly, Gewirth's argument recognises that the impulse of sociability by being merely contingent is not sufficient or even necessary in providing the foundation for universal morality. What is required, as Stage (3) of the argument for oikeiosis brings out, is the recognition that virtue is not only a matter of acting correctly but more importantly acting for the right reasons; not merely a matter of acting out of a social instinct or impulse, that at best can only be contingent and variable, but a matter of acting rationally.

The Stoic switch from the pursuit of natural aims motivated by both the impulses for self-preservation and sociability, to an exclusive interest in rational agreement with nature that marks the crucial Stoic transition to virtue is, however, not clearly demonstrated by argument. We are led to understand that the belief that nature is rational and orderly together with the belief that one's own individual nature reflects and is part of that natural rational order, are together sufficient to rationally convince a person that he ought to live in agreement with nature that, in turn, requires virtuous conduct. However, Gisela Striker correctly points out (Striker 1996, 225–31) that the dual realisations that nature is rational and orderly and that one's own individual nature reflects and is part of that order, are not sufficiently compelling to support the transition from natural and social impulses to the rational "impulse" of acting and living virtuously in accordance with reason. For a person can ascent to the truth of both those propositions without being rationally compelled to accept the view that a good life is a life of virtue. For one can accept that nature is both rational and orderly and that one's own nature reflects that rational order without being rationally compelled to accept that that entails living virtuously and that living in agreement with nature is ethically good.

Stage (II) in Gewirth's argument for the PGC closes that logical gap in the Stoic argument for oikeiosis by demonstrating that an agent is logically compelled on the basis of his own rational nature (rational purposive agency) to recognise and accept the transition from a self-regarding prudential perspective to an other-regarding moral perspective, on pain of self-contradiction (Gewirth 1978; 1996; Spence 2006). If my analysis is right, Gewirth's argument provides the Stoic argument for oikeiosis with the missing rational lacuna that closes the logical gap between the natural and social impulses in Stage (2) of that argument, and the transition to virtuous conduct, required by reason, in Stage (3).

(III) As mentioned earlier, the transition to constitutive virtue and Gewirthian neo-Stoic sagehood is provided by Gewirth's indirect applications of the PGC through his model of self-fulfillment. That final transition from developmental to constitutive virtue and ideal self-fulfillment parallels the Stoic transition from partial and imperfect virtue to complete and perfect virtue, a transition that marks Stoic sagehood. In conclusion, Gewirth's argument for the PGC mirrors, at least schematically if not in detail, the developmental structure of the Stoic argument for oikeiosis. In addition, the rational support provided indirectly for the argument for the PGC for both developmental and constitutive virtue as well as the arguments advanced by Gewirth on the basis of the argument for the PGC in support of his model of self-fulfillment, mirror and parallel the various Stoic arguments and doctrines for virtue and eudaimonia. Central to both Gewirth's argument for self-fulfillment and Stoic eudaimonia are the key elements of rationality, virtue and agreement with nature, which in Gewirthian terms translates to agreement with the logical implications of rational purposive agency and personhood.

CONCLUSION

The comparative analysis between Stoic philosophy and a neo-Stoic philosophy, I have attributed to Alan Gewirth based on my reconstruction of his theory of *self-fulfillment* and generally his rationalist ethics, has demonstrated a clear and close parallel between Gewirthian and Stoic ethics on the one hand, and a close parallel between Gewirthian ideal self-fulfillment (second-order self-fulfillment) and Stoic eudaimonia. In conclusion, Gewirth's ethical theory, including his model of self-fulfillment, bears a strong and striking similarity, methodically and contextually, to Stoic ethics and Stoic eudaimonia. Insofar as my comparative analysis is correct, Gewirth's ethical theory, including his model of self-fulfillment, is paradigmatically and essentially neo-Stoic. I believe this is a significant and welcome result for both contemporary ethics and Stoic scholarship. In the next chapter, we examine how Stoic and neo-Stoic philosophy applies specifically to technology.

NOTES

1. A. A. Long, *Hellenistic Philosophy: Stoics, Epicureans, Skeptics* (Berkeley and Los Angeles: University of California Press, 1986, 2nd-edn), Striker, *Essays on Hellenistic Epistemology and Ethics* (Cambridge: Cambridge University Press, 1996), J. Annas, *The Morality of Happiness* (New York: Oxford University Press, 1993), M. C. Nussbaum, *The Therapy of Desire* (Princeton, NJ: Princeton University Press, 1994), and L.C. Becker, *A New Stoicism* (Princeton, NJ: Princeton University Press, 1998); John Sellars, *Lessons in Stoicism: What Ancient Philosophers Teach Us about How to Live* (Penguin Books, 2020); John Sellars, *Stoicism* (Berkeley University California Press, 2020); Brad Inwood, *Stoicism: A Very Short Introduction* (Oxford:

Oxford University Press, 2018); Pierre Hadot, *The Present Is Our Happiness* (Palo Alto: Stanford University, 2011); Pierre Hadot, *The Inner Citadel: The Meditations of Marcus Aurelius* (Cambridge, Massachusetts: Harvard University Press, 2001); Pierre Hadot, *Philosophy as a Way of Life* (Oxford: Blackwell Publishing, 1995), Robertson, Donald. 2019. How to Think Like and Emperor: The Stoic Philosophy of Marcus Aurelius. New York: Martin's Griffin, Macmillan Press.

2. In 1997, Edward H. Spence founded the *Philosophy Plays Project*. Since then to the present time (2021) he has produced and directed, in the spirit of Platonic Symposia, a program of public philosophy in Australia through the medium of philosophy plays, that combine philosophical talks by academic philosophers with dramatic performances by actors. Several of the philosophy plays written by Spence have been performed in various Arts Festivals in Australia. His philosophy play, *The Philosophy of Love: Love in the Age of Terror,* was performed at the Sydney Opera House in 2004; his Stoic philosophy play, *Zeno's Secret: How to be Happy in an Unhappy World*, was performed at the Factory Theatre in 2018 for the Greek Festival of Sydney and in 2021 his philosophy play *The Last Freedom: Algorithms and Ataraxia,* a play based on the theme of this book on Stoic Philosophy and Technology, will be performed for the Greek Festival of Sydney at the Art Hellenic Theatre. The ultimate aim of the project is to cultivate and socialize philosophy by raising public awareness of its practical application and usefulness as a "way of life" in helping us, as individuals and collectively as communities and society generally, to think more critically and reflectively about how we live our lives. In Martha Nussbaum's apt phrase, taken from the title of her book, it is a way of "cultivating humanity" (Nussbaum 1997).

3. For Stoicism in public philosophy and professional ethics see, E. H. Spence (2004), "Philosophy Plays: A Neo-Stoic Method for Teaching Ethics," *Teaching Ethics*, vol. 5, no. 1 (2004), 41–57 and E. H. Spence "The Theatre of Philosophy: A Neo-Socratic Method for Performing Public Philosophy," in Bull, Gordon (Ed.), *Creativity: brain – mind – body: A view into the future of Australian Art and Design Schools, the 2011 Australian Council of University Art and Design Schools (ACUADS) Conference*, the Australian National University, the University of Canberra, and the Canberra Institute of Technology, 21–23 September 2011 (2012).

4. Though as we shall see, Epicureanism, when examined more closely, is not that dissimilar from Stoicism in its fundamental ethical principles of living a good life for the ultimate purpose of attaining eudaimonia.

5. See Spence, E. *Ethics Within Reason: A Neo–Gewithian Approach.* (Lanham, Maryland: Lexington Books, a division of Rowman and Littlefield, 2006), pp. 393–442.

6. When in conversation with Gewirth (Chicago, March 2001), I raised the issue of the striking similarity between his ethics and that of the Stoics, he expressed both interest and surprise. I hope to show that the surprise for Gewirth should be a pleasant one and the interest is both warranted and justified.

7. For a comparison of Kant's and Gewirth's ethical theories, see Spence E., A Commentary on Deryck Beyleveld's "Gewirth and Kant on Justifying the Supreme Principle of Morality." In Michael Boylan (ed.), *Gewirth, Critical Essays On Action, Rationality, And Community*. Maryland: Rowman and Littlefield Publishers, Inc. pp 119–124. ISBN: 9780847692590.

8. See especially his *Self-fulfillment* (Princeton, NJ: Princeton University Press, 1998). Although references to Aristotle abound, and the ones to Kant are not infrequent, the references to the Stoics are almost conspicuously scarce, if not entirely absent.

9. An earlier longer version of this chapter can be found in Spence, E. (2006) *Ethics with Reason: A Neo-Gewirthian Approach* (Chapter 10).

10. G. Striker, "Stoicism" in Lawrence C. Becker and Charlotte B. Becker, (eds.), *Encyclopedia of Ethics* (New York: Garland Publishing, 1992), pp. 1208–1213.

11. These sources are: J. Annas, *The Morality of Happiness* (New York: Oxford University Press, 1993); J. Annas, *Platonic Ethics, Old and New* (Ithaca: Cornell University Press, 1999); L. C. Becker, *A New Stoicism* (Princeton, NJ: Princeton University Press, 1998); A. A. Long, *Hellenistic Philosophy – Stoics, Epicureans, Sceptics* (Berkeley: University of California Press, 1986, 2nd Edition); G. Striker, *Essays On Hellenistic Epistemology and Ethics* (Cambridge: Cambridge University Press, 1996); G. Striker, "Stoicism" in Lawrence C. Becker and Charlotte B. Becker (eds.), *Encyclopedia of Ethics* (New York: Garland Publishing, 1992).

12. General readings that inform this chapter are the following: Epictetus *Discourses and Selected Writings*: Marcus Aurelius *Meditations*; Seneca's *Letters from a Stoic*; Pierre Hadot's *The Inner Citadel: The Meditations of Marcus Aurelius*; and *Philosophy as a Way of Life*; *Stoicism* by John Sellars.

13. Long, *Hellenistic Philosophy*, 194

14. For various scholarly views on the "final goal" or "final good" od Stoic philosophy, see Striker, "Stoicism" 1209, and *Essays*, 222–24; Long, *Hellenistic Philosophy*, 194, 199; and Becker, *A New Stoicism*, 133–34.

15. This rational and moral development is described in detail by one of the most central doctrines of Stoic ethics, that of *oikeiosis*.

16. Gewirth, *Self-fulfillment*, 46, 59, 182–89, 199, 223.

17. In *Self-fulfillment*, Gewirth distinguishes between self-fulfillment as *aspiration fulfilment* and as *capacity fulfilment*. The former he identifies as the satisfaction of one's deepest desires, the latter as the process and goal of making the best of oneself.

18. Long, Hellenistic Philosophy, 174, 189–96, 211, 215–16; Striker, *Essays*, 219, 222–31.

19. Gewirth, *Self-fulfillment*, 69–70, 86, 91–92, 97–98, 101, 109–10, 222.

20. Striker, "Stoicism," 2011, and *Essays*, 182, 194, 224; Becker, *A New Stoicism*, 127, 138–49, 150–56; Long, *Hellenistic Philosophy*, 197,234.

21. Gewirth, *Self-fulfillment*, 14–16;18;25;50;62;85;107;182–89.

22. When comparing Gewirthian self-fulfillment to Stoic eudaimonia, let it be understood that I am referring specifically to Gewirth's *second-order*, or *ideal* notion *of self-fulfillment*.

23. Stoics: Striker, "Stoicism" 209–10, and *Essays*, 181–82, 240–43, 255, 279, 288; Becker, *A New Stoicism*, 118–22; Long, *Hellenistic Philosophy* 192, 197–205; Gewirth: *Self-Fulfillment*, 14, 56–58, 68–9, 79, 87–90, 114–15, 122–23, 126–27, 131, 134–35, 175–77.

24. Stoics. Striker, "Stoicism" 2011 and *Essays*, 182, 194, 224; Becker, *A New Stoicism*, 127–29, 138–49, 150–56; Long, *Hellenistic Philosophy*, 197, 134.

25. In her review of Martha Nussbaum's *Cultivating Humanity: A Classical Defense of Reform in Liberal Education* (1997), Marilyn Friedman points out that the "idea of love for all humanity," is mentioned by Nussbaum "no less than fourteen times" (*Cultivating Humanity*, pages 6, 7, 13, 14, 36, 61, 64, 67, 72, 84, 103, 222, 259, 292) – see Marilyn Friedman, "Educating for World Citizenship" *Ethics* (April 2000): 589. Nussbaum's "love for all humanity" seems essentially similar to the 'love of the Good' which, as the primary motivation for doing what is right, I attribute to both the Stoic and the neo-Stoic.

26. In my analysis, I have followed closely Laurence C Becker's insightful neo-Stoic and Stoic analysis on the sufficiency thesis of Stoic eudaimonia (ANS:108–158)

27. For a detailed discussion of Stoic *Oikeiosis* and its structural similarity with a similar concept used by Gewirth in his ethical theory, see E. Spence, *Ethics Within Reason: A Neo-Gewithian Approach* (Maryland: Lexington Books, a division of Rowman and Littlefield, 2006), pp. 433–442.

28. To remind us of what was said earlier in regards to virtue, Gewirth makes a distinction between d*evelopmental* and *constitutive* virtue. D*evelopmental* virtue is what enables one to become ethically a good person. Its function is primarily instrumental as physical fitness is for health. However, once one becomes ethically good, in the sense that one becomes disposed to always act out of virtue, which at this stage of one's development has become identical with one's rational purposive agency, virtue becomes *constitutive* of that person's ethical goodness. To use the same previous analogy, one's physical fitness is no longer merely instrumental for becoming healthy, but it becomes, once one attains and maintains good health, constitutive of one's health and general lifestyle. When constitutive, virtue, like physical fitness, is its own reward – it becomes together with moral goodness an end in itself and not merely an instrumental means for ethical conduct.

29. The necessity of purposive agency as a prerequisite for Stoic, and in particular, neo-Stoic eudaimonia or happiness is also alluded to by M. Nussbaum, in *Symposium on Cosmopolitanism*, "Duties of Justice, Duties of Material Aid: Cicero's Problematic Legacy," *The Journal of Political Philosophy*, vol. 8, no. 2 (2000), 191.

30. Stoics: Becker, *A New Stoicism*, 56; 74; 90; 93–95; Long, *Hellenistic Philosophy*, 172–74; 185–88; Striker, "Stoicism," 1209–1210, and *Essays*, 178;181; 219; 225–26; 250–51; 253–54; 257–261; 267; 281–297.

31. *Self-fulfillment*, 45;60;65;86;95–96;112;123–35;222–23;226.

REFERENCES

Annas, Julia. 1999. *Platonic Ethics, Old and New*. Ithaca: Cornell University Press.
———. 1993. *The Morality of Happiness*. New York: Oxford University Press.
Becker, Laurence. 1998. *A New Stoicism*. Princeton: Princeton University Press.
Cicero, *Tusculan Disputations* (Loeb Classical Library) 2nd Edition, translated by J. E. King. Cambridge, MA, and London: Harvard University Press, 1927.

Epictetus. 2018. *Epictetus: Discourses and Selected Writings*, translated and edited by Robert Dobbin. London: Penguin Classics.
Friedman, Marilyn. 2000. "Educating for World Citizenship," *Ethics*, April, 589.
Gewirth, Alan. 1998. *Self–fulfillment*. Princeton: Princeton University Press.
———. 1996. *The Community of Rights*. Chicago: University of Chicago Press.
———. 1978. *Reason and Morality*. Chicago: University of Chicago Press.
Hadot, Pierre.2001. *The Inner Citadel: The Meditations of Marcus Aurelius*. Cambridge, MA: Harvard University Press.
———. 1995. *Philosophy as a Way of Life*. Oxford: Blackwell Publishing, 273.
Laertius, Diogenes. 1931. *Lives of Eminent Philosophers Books 6-10*. Cambridge, MA: Harvard University Press.
Long, A. A. 1986, *Hellenistic Philosophy – Stoics, Epicureans, Sceptics*. 2nd Edition. Berkeley: University of California Press.
Nussbaum, Martha. 2000. "Duties of Justice, Duties of Material Aid: Cicero's Problematic Legacy." In *Symposium on Cosmopolitanism. The Journal of Political Philosophy*, vol. 8, no. 2, 191.
———. 1997. *Cultivating Humanity: A Classical Defense of Reform in Liberal Education*. Cambridge, MA, and London: Harvard University Press.
———. 1994. *The Therapy of Desire*. Princeton, NJ: Princeton University Press.
Spence, Edward H. 2011. "The Theatre of Philosophy: A Neo–Socratic Method for Performing Public Philosophy', In Bull, Gordon (Ed.), *Creativity: brain – mind – body: A view into the future of Australian Art and Design Schools, the 2011 Australian Council of University Art and Design Schools (ACUADS) Conference*, the Australian National University, the University of Canberra, and the Canberra Institute of Technology, 21–23 September.
———. 2007. "Positive Rights and the Cosmopolitan Community: Right–Centered Foundations for Global Ethics." *Journal of Global Ethics*, vol. 3, no. 02, July/August (2007). New York: Taylor and Francis.
———. 2006. *Ethics Within Reason: A Neo–Gewithian Approach*. Lanham, Maryland: Lexington Books, a division of Rowman and Littlefield, 433–442.
———. 2004."Philosophy Plays: A Neo–Stoic Method for Teaching Ethics." Utah, USA: *Teaching Ethics* vol. 5, no. 1, 41–57.
———. "A Commentary on Deryck Beyleveld's "Gewirth and Kant on Justifying the Supreme Principle of Morality." In Michael Boylan (ed.), *Gewirth, Critical Essays On Action, Rationality, and Community*. Maryland: Rowman and Littlefield Publishers, Inc. pp 119–124. ISBN: 9780847692590.
Striker, Gisela. 1996. *Essays on Hellenistic Epistemology and Ethics*. Cambridge: Cambridge University Press.
———. 1992. "Stoicism" In Lawrence C. Becker and Charlotte B. Becker (Eds.), *Encyclopedia of Ethics*. New York: Garland Publishing, 1208–1213.

Chapter 4

Application of Stoic Philosophy to Technology

O my friend, why do you who are a citizen of the great and mighty and wise city of Athens, care so much about laying up the greatest amount of money and honour and reputation, and so little about wisdom and truth . . . Are you not ashamed of this?

—Plato, *Apology*, 1978.

INTRODUCTION

This chapter examines how Stoic and neo-Stoic philosophy applies to technology, and specifically information communication technologies (ICTs) and Artificial Intelligence (AI) technologies. Unless otherwise stated for convenience, I will refer to these technologies collectively as "technology." The chapter comprises three parts: (1) the core problem of Stoic philosophy and technology; (2) the core principles of Stoic philosophy and how they apply to technology; and (3) why Stoic philosophy is of relevance and importance to technology.

Let me begin by clarifying, that the problem of technology is not technology as such, but the use of technology by the Big Tech companies, such as Facebook and Google, for their primary financial gain to the detriment of the users' well-being and society generally. The recent scandal of Cambridge Analytica and Facebook involving breach of privacy and lack of consent of users' data is a case in point. That case and its normative implications and impact on society will be examined in chapter 6.

As we saw in chapter 2, given that any technology cannot be good for itself or the good of the Big Tech companies that manage them, but good for humanity overall, any instrumental positive value of such technologies must not be at the cost of their negative eudaimonic impact on society. Ultimately, the core problem of technology is not technical to be solved by engineers but a moral problem to be solved collectively by society and the global community.

The overarching rationale for the application of Stoic philosophy to evaluate the impact of technology on society is that Stoic philosophy as a way of life with its focus on the attainment of eudaimonia or well-being is conceptually and practically relevant and important to the eudaimonic assessment of technology. The Dual Obligation Information-Wisdom Theory (DOIT-Wisdom) designed to enable the operationalisation of the application of Stoic ethics to technology forms a central feature of the application of a neo-Stoic philosophy to technology, a topic we examined in chapter 3. The DOIT-Wisdom model will be examined in more detail, in chapter 5.

THE CORE PROBLEM OF STOIC PHILOSOPHY AND TECHNOLOGY

In both Stoic philosophy and technology, the key problem and question, and the focus of this chapter, and generally of this book, is the question of who is in control. Is it **Us**, as individuals and collectively as a society and the global community generally, or **Technology**? The chapter will demonstrate that the answer provided by the Stoics to the question of who is in control of our lives, *Us*, or *Circumstance*, signals and motivates an answer to the question of who is in control of technology? Is it *Us*, as individuals and collectively as society, or the *Big Tech companies* who control that technology, and specifically, ICTs and AI technologies?

The Problem of Stoic Philosophy: Who Is in Control? Us or Fortune?

Stoic philosophy sets out to answer the key meta-philosophical question of how to be happy in an unhappy world, despite the slings and arrows of outrageous fortune, a key feature Stoicism shares with Buddhism. The answer Stoicism provides is methodologically and logically ingenious, at once theoretical and practical. Perceiving philosophy to be a therapy of the mind, a *Techne Viou or the Art of Living*, the Stoics observed that there are things within our control and things outside our control.

According to the Stoics, our main concern should be with developing our virtues, such as courage, moderation, justice, prudence or practical wisdom, empathy and solidarity, as virtue is the only inherent good, which provides both the necessary and sufficient practical means for the attainment of eudaimonia. To avoid unhappiness, we must do two things: control things that are within our control (our beliefs, judgements, desires, attitudes and volition or *prohairesis*) and be indifferent to things that are not in our control, or at least not in our full control, that is, things external to us, such as health, relationships, employment, wealth, power, death, and so on. As rational beings, we should therefore perfect our characters through living a virtuous life in accordance with nature because it can bring us both *ataraxia* (tranquillity); *autarkia* (self-sufficiency) and *eudaimonia* (happiness or well-being). A happiness that is within our control, just because it depends on our own rational volition and judgements that are within our control.

As we saw in chapter 3, an important feature of Stoic eudaimonia is its intrinsic relation to virtue. For virtue as the only inherent good is not only the essential means for the attainment of eudaimonia but also constitutive of it. In being the ultimate goal of a good life, virtue is therefore both the means and the end for a eudaimonic or happy life, or at least, not an unhappy life. Eudaimonia, which in ancient Greek means having within us a good daemon or spirit, is up to us, as it lies within us and it is, therefore, within our control. Interestingly, the modern Greek word for happiness, *eutihia* (pronounced in modern Greek *eftihia*) that means having good fortune, is not within our control, as it places happiness outside our control, since fortune or circumstance, good or bad, lies outside our control.

The Core Problem of Technology: Who Is in Control? Us or Technology?

The core problem of technology, and specifically with regard to this chapter, ICTs, and AI technologies, is that those technologies are not within our control but in the control of the Big Tech companies, such as Facebook and Google, as manifested in practice by the systemic surveillance of users' information by opaque and inscrutable algorithms, for the financial gain of Big Tech companies, dictated by their Business Model. That problem has been widely investigated by many researchers in recent years, notably Shoshanna Zuboff (2019) and a number of other researchers.[1]

The primary purpose of this chapter is to show how Stoic philosophy is both relevant and effective in theorising that problem, as well as practically responding to it for its solution. Having identified the parallel control problems of Stoic philosophy and technology, the chapter will show how Stoic philosophy is both theoretically relevant and practical to technology through its core principles we examined in detail in chapter 3. To that end, this chapter will summarise in outline those core principles and show how they apply to technology, and in particular ICTs and AI technologies.

THE PRINCIPLES OF STOIC PHILOSOPHY AND THEIR APPLICATION TO TECHNOLOGY

Principle 1: Concern Only for Things within Our Control

According to Epictetus (2018, 221),

> Our opinions are up to us, and our impulses, desires, aversions – in short, whatever is our doing. Our bodies are not up to us, nor our possessions, our reputations, or our public offices, or, that is, whatever in not our doing . . . So remember, if you think that things naturally enslaved are free or that things not your own are your own, you will be thwarted, miserable, and upset, and will blame both gods and men. (*Enchiridion,* Ch.1:1–3)

As the issue of **Control** is a core problem that arises in both Stoic philosophy and technology, Stoic philosophy is well placed to provide a theoretical and practical evaluation, as well as motivation to the solution of the control problem of technology.

If core aspects of technology are outside our control, there are at least two options open to us within Stoic methodology:

> The first option is that since accuracy, reliability, trustworthiness, transparency, accountability, privacy, the inscrutability, monopoly and the asymmetry of information between Big Tech and us, as users of those technologies, and ultimately, the control of technology itself, is beyond our control, we should ignore such matters, or at least treat them as preferred indifferents not required for our well-being, since we cannot control them.
>
> The second option is precisely because of the harmful impact those technologies have on the epistemic, ethical and eudaimonic features over most aspects of our lives, we should endeavour to bring those technologies under society's control, since our collective well-being is at risk if we don't.

A key underlying assumption of the Stoic notion of *control* is inner mental freedom and autonomous rational agency. However, the psychologically persuasive technologies used by the Big Tech companies to inculcate users to spend maximum time on Big Tech's digital platforms (since more time spent is more money earned by them) are designed to undermine the users' rational autonomous agency for the primary financial gain of those Big Tech companies. Which, like concealed and inscrutable puppeteers want to make us dance to their Faustian-Midas's tune,[2] just as the prisoners in Plato's *Allegory of the Cave*. In chapters 6–9, we shall examine more closely how Big Tech companies, such as Facebook and Google, using their surveillance Business Models through the application of opaque persuasive algorithms, undermine the users' rational autonomous agency for the primary financial gain of those companies: by collecting and selling users' data to advertisers and other information contractors, without the users' explicit knowledge and consent.

Principle 2: The Final Goal

The Hellenistic philosophers, including the Stoics, claimed that *eudaimonia* or happiness was the ultimate goal of a good life, though each of the Hellenistic schools identified happiness with something different: The Epicureans, identified it with pleasure and the avoidance of pain, the Pyrrhonian Sceptics with suspension of all judgement as the only sure way to achieve *ataraxia* or tranquillity, and the Stoics identified eudaimonia exclusively with a life of virtue, which, in turn, they identified with a life in agreement with nature. It is, so to speak, our "nature" to be rational, and it is our rational nature that leads us through various intermediary stages, to a virtuous and ethical life that ultimately results in the attainment of eudaimonia or happiness.

The emphasis of the Stoics on the attainment of eudaimonia through virtue is significant for examining the impact of technology on our individual and collective well-being, and how we can respond to it. It highlights why technology cannot be good for itself or for the Big Tech companies alone that manage it, but for the common good of all humanity, which was the topic we examined in chapter 2.

Principle 3: The Life of Virtue and the Life of Wisdom

As we saw in chapter 3, the Stoic definition of virtue is as follows:

(I) Virtue is not defined by the consequences in the world which it succeeds in promoting, but by a pattern of behaviour that follows necessarily from a disposition perfectly in tune with Nature's rationality.
(II) The right thing to do is that which accords with virtue, and this is equivalent to saying that it accords with the nature of a perfect rational being.
(III) Virtue accords with Nature in the sense that it is the special function or goal of a rational being to be virtuous. (Long1986, 192)

Associated with the Stoic principle of a virtuous life, is the principle of Wisdom. The ultimate object of philosophy according to the Stoics, and the ancient Greek philosophers generally, beginning with Socrates, is to teach us not just knowledge, but wisdom. Wisdom, therefore, for the Stoics is a way of a life of virtue that brings peace of mind, inner freedom, happiness, and a cosmic consciousness. By "cosmic consciousness," the Stoics understood that quality, as universal reason, which permeates the whole universe and by virtue of which we are all, integrated parts of the cosmos. As Pierre Hadot tells us (1995, 273),

> The exercise of wisdom entails a cosmic dimension. Whereas the average person has lost touch with the world, and does not see the world qua world, but rather treats the world as a means of satisfying his desires, the [stoic] sage never ceases to have the whole constantly present to mind. He thinks and acts within a cosmic perspective. He has a feeling of belonging to a whole, which goes beyond the limits of his individuality.

Significantly, more than information, wisdom requires transformation. To become wise, one requires not only to be informed but to become transformed, through living virtuously, by the practice of philosophy as a way of life. Given the unwise ways in which we use technology, including our uncritical acceptance of the control Big Tech companies exert over all facets of our lives, we should use information, more virtuously and wisely for the collective good and well-being of ourselves and society. Wisdom, as *meta-technology of the mind* (Spence 2011) can guide us on how to control technology and our use of it, for our common good. The topic of wisdom will be explored in more detail in chapter 5.

Principle 4: Cosmopolitanism

When anyone asked him where he came from, he said, "I am a citizen of the world" (Laertius 1925, Chapter 2, 22–84).

Closely associated with the notion of cosmic consciousness is the Stoic cosmopolitan perspective. It involves living empathetically and in solidarity for the common good of the world community and acting in accordance with global social justice for our collective well-being. According to the Stoics, philosophy entails a community engagement. For the Stoics, however, the polis was the cosmopolis, not the city state, or individual nations, but the whole world.

The threat to the natural environment through climate change and the current COVID-19 pandemic renders the ethical lifestyle recommended by the Stoics, which respects the ethical rights of all individuals as citizens of the world, and also respects both the natural and digital environments, not only desirable, but necessary. In the connected technological age in which we live, Stoic cosmopolitanism as a way of life for the attainment of well-being is relevant and important for enabling us to mount a collective response to

Big Tech's surveillance practices that systemically undermine our collective well-being. Since we cannot mount such a response as individuals, our response, just as our response to a solution to climate change, requires a global collective action, that incorporates, citizens, governments and their associated democratic institutions and regulatory authorities, as well as the Big Tech companies themselves. Stoic cosmopolitanism provides both the theoretical framework and practical motivation for such a collective shared response.

Principle 5: A Life in Agreement with Nature

As Cicero explains, the value of anything in Stoicism is defined by reference to Nature (Long 1986, 198).

An important and core principle of Stoicism is living in accordance with nature. Eudaimonia, according to the Stoics, consists in living a virtuous life and a virtuous life is one in agreement with nature, which they identified with reason or rational spirit that permeates the entire universe. Since nature was identified with universal reason, the Stoics identified a life in agreement with nature with the perfection of human reason. Moreover, it is only by living a rational and virtuous life that human beings can truly become happy. For the Stoics, there is an integral relationship between virtue, rational agency, nature and well-being, all of which relate to living a good life for the attainment of eudaimonia.

Since, the Stoic notion of nature, under the influence of Heraclitus, was conceived to be dynamic and not static, we can extend the Stoic notion of nature to include not only the natural environment but also crucially the digital environment. Applying the Stoic principle of living in agreement with nature we can say that just as we have a collective global responsibility in respecting and safeguarding the integrity of the natural environment (biosphere), we have an equal responsibility with regard to the digital environment (infosphere). Whereas climate change pollutes and corrupts the biosphere, the surveillance practices of the Big Tech corrupt the infosphere. Both are a form of corruption that undermine both our individual and collective global well-being. In chapter 6, we shall examine the notion of media corruption with regard to some of the information and communication practices of Big Tech companies, such as Facebook and Google.

Principle 6: Oikeiosis

Related to the issue of transformation through wisdom examined earlier, is the principle of *oikeiosis*. As we saw in chapter 3, the transition from developmental to constitutive virtue, from imperfect and incomplete virtue to perfect and complete virtue, requires a kind of ethical transformation. In an insightful passage, Julia Annas describes quite clearly the kind of ethical transformation that the Socratic theory of happiness requires and which by extension applies equally to Stoic philosophy, as the Stoics considered Socrates to be their earlier role model (Annas1999, 50–51):

> As a theory of happiness [the Socratic theory of happiness in Plato], it is very demanding. You won't have got your account of happiness right until you achieve the right perspective, and that will be utterly different from your initial standpoint. Becoming virtuous requires *transforming* [emphasis added] your values and priorities; nothing short of this will do. All your valuations of conventional goods and evils will turn out to be wrong . . . we are used to moral theories that are demanding, but it is not easy for us to conceive of such a theory which is a theory of *happiness*, rather than something else. But this is not an argument against a theory like Plato's. It shows merely that it takes a long intellectual journey away from common sense and conventional beliefs to achieve the wisdom of the virtuous person and thus to have your whole life, and your conception of happiness transformed.

Alan Gewirth also recognises and acknowledges the transformative aspect of his concept of ideal self-fulfillment. As he tells us (Gewirth 1998, 86), "to recognize and accept these requirements in one's thought, and action may involve a kind of *self-transformation* [emphasis added] whereby one becomes more fully a moral and social being."

It is the realisation of the practical necessity for such transformation and the need for providing a conceptual account of how such a transformation *can practically* take place, that the Stoics developed their theory of *oikeiosis*; a theory considered to be one of the most crucial and fundamental doctrines of Stoic ethics. The theory purports to show how a person can progressively develop, because he is naturally conditioned and thus capable of doing so, from the state of having and being motivated by purely hedonistic desires as a child, to the state of acquiring, over a time period of biological, psychological, social, but more importantly, rational development, an ethical perspective.

As we saw in chapter 3, the four stages comprising the central developmental stages in the Stoic doctrine of oikeiosis are: the impulse for *self-preservation*; the impulse for *sociability* and the recognition that one should not merely act appropriately towards oneself and others on the basis of "natural impulses" alone, but more importantly should do so on the basis of their *rationality*. And finally, the disposition and the practice of always acting according to one's perfected reason and virtue in agreement with nature, results in the attainment of *eudaimonia*. For this is in accordance with one's own rational nature that reflects and is part of, the rational order of nature, and the universe. Acting rationally is appropriate because it is in agreement with nature and agreement with nature is necessary if one is to attain eudaimonia, which is the ultimate goal of a good life (see Principle The Final Goal above).

As we shall see in chapter 5, the *DOIT-Wisdom* which advances a neo-Stoic philosophy based on a reconstruction of Gewirth's theory of *self-fulfillment*, examined in chapter 3, aligns with the Stoic notion of *oikeiosis*. Gewirth's argument for the *principle of generic consistency (PGC)*, which supports universal human rights for freedom and well-being, mirrors, at least schematically if not in detail, the developmental structure of the Stoic argument for oikeiosis.

In addition, the rational support provided indirectly for the argument for the PGC for both *developmental* and *constitutive* virtue as well as the arguments advanced by Gewirth on the basis of the argument for the PGC in support of his model of self-fulfillment, mirror and parallel the Stoic arguments and doctrines for virtue and eudaimonia. Central to both Gewirth's argument for self-fulfillment, and Stoic eudaimonia are the key elements of rationality, virtue, and agreement with nature, which in Gewirthian terms translates to agreement with the logical implications of rational purposive agency and personhood. And as we saw earlier in examining the problem of control regarding technology, the problem is how Big Tech companies, such as Google and Facebook, target our rational autonomy and agency through their systemic persuasive opaque algorithms for their own self-centred financial gain, and do so without any transparency and accountability.

Applying the principle of oikeiosis to technology further emphasises the urgent need why we must transform and no longer accept what is clearly harmful to us. Namely, the wilful deception and manipulation of our *autonomy* and *privacy*[3] by Big Tech companies. In later chapters of this book, and especially in chapters 5, and 7–9, we shall examine in more detail how the concepts of autonomy and privacy are closely interrelated and both essential for our fundamental rights to freedom and well-being as well as our dignity, as individuals and collectively as society.

In the introduction of her important book, *Privacy is Power*, Carissa Véliz (2020, 1) sounds a siren warning of how Big Tech violates and usurps our privacy, thus undermining our individual and collective well-being:

> They are watching us. They know I'm writing these words. They know you are reading them. Governments and hundreds of corporations are spying on you and me, and everyone we know. Every minute of every day. They rack and record all they can: our location, our communications, our internet searches, our biometric information, our social relations, our purchases, and much more. They want to know who we are, what we think, where we hurt. They want to predict and

influence our behaviour. They have too much power. Their power stems from us, from you, from your data. *It's time to take back control* [emphasis added]. Reclaiming privacy is the only way we can regain control of our lives and our societies.

Taking back control of technology by taking control of our autonomy and privacy, as individuals and collectively as society, is also the intended clarion call of this book, guided by the normative and eudaimonic conceptual framework and rational of Stoic and neo-Stoic philosophy. For as this chapter has shown the solution provided by Stoic and neo-Stoic philosophy to the control problem of who is in control of our lives, Us or Circumstance, offers also a response to the control problem of technology, that is, who is in control of technology, Us or Big Tech. It is a matter of necessity, therefore, to mount a collective cosmopolitan response to Big Tech, as the condition for our well-being and ultimately our survival as free and autonomous human beings, depends on it. As we shall see in chapter 9, when examining in more detail the control problem of AI technology, the survival of humanity itself, depends on it.

Principle 7: Philosophy as a Way of Life

> Empty is that philosopher's argument by which no human suffering is therapeutically treated. For just as there is no use in a medical art that does not cast out the sicknesses of the body, so too there is no use in philosophy, unless it casts out the suffering of the soul. (Epicurus, quoted in Nussbaum1994, 13)

An important aspect of Stoic philosophy, as well as the other Schools of Hellenistic philosophy, and in particular, Epicurean philosophy as the quote from Epicurus above indicates, was that philosophy had to be conducted as a way of life or the art of living. Philosophy, as we saw in chapter 3, can be viewed as the art of living in two ways:

(a) As a craft or art which once learned can enable one to live a good and fulfilling life on a daily basis and
(b) The perfection of one's human nature – becoming the best of one's kind – humankind. For the Stoics, the perfection of one's human nature meant the perfection of one's rational nature. For according to the Stoics, the essence of human nature and nature generally, the whole universe in fact, was reason. Thus, in perfecting one's rational nature one became at once the object and the subject of art, that is, the art of life. This was the highest form of art, in perfecting one's rational nature through the practice of philosophy. In the neo-Stoic theory, I based on my reconstruction

of Gewirth's notion of self-fulfillment, a life in agreement with nature is a life in agreement with the perfection of one's rational purposive agency, which, in turn, is in agreement with universal morality, which is essential for a good and self-fulfilled life. In contrast to the Stoic conception of universal nature as divine, however, Gewirth's notion of "nature" is and idealised *human nature*, understood as rational purposive agency applied to the aim of becoming the best oneself as a human being, which as we saw earlier requires nothing less than a transformation through the exercise of wisdom. Wisdom, as a form of experiential meta-knowledge of how to live a life of virtue for the attainment of eudaimonia, requires philosophy to be lived in practice as a way of life and not merely an intellectual exercise of thought alone. As Pierre Hadot tells us (1995, Chapter 11),

> Philosophy as a way of life for the Stoics was, A mode of existing-in-the-world, which had to be practiced at each instant, and the goal of which was to transform the whole of the individual's life ... the word *philo-sophia* – the love of wisdom – was enough to express this conception of philosophy.

Significantly, the application of Stoic and neo-Stoic philosophy as a way of life in addressing and responding to the problem of technology examined above, namely the problem of control by the Big Tech companies over all aspects of our lives, is well suited to a global digital network environment: both for its identification and eudaimonic evaluation of its impact on society, as well as the enablement of its solution, through the shared collective normative response by users of technology and society and the global community generally. For just as Big Tech can use the user's information to the detriment of their users' well-being, the users themselves have the capacity and ability to apply their shared knowledge of the control problem of technology to turn the tables on Big Tech. That is because the users can at a stroke put a significant dent in the Big Tech's business model of clickbait and consequently their billion-dollar revenue stream, obliging the Big Tech companies to take appropriate action to remedy their dis-demonic manipulative practices against the users' legitimate normative claims.

For just imagine for a moment what would happen if the billions of Facebook's and Google users worldwide went on strike and stopped using their platform services, even for a day. Both companies would lose a lot of money and their surveillance-based business model would shudder. Of course, the practicality of such shared collective action may prove difficult but as they say, where there is a will, there is a way. Stoic philosophy is at least theoretically and practically well-equipped to motivate and provide both the will and the way.

CONCLUSION: WHY THE APPLICATION OF STOIC PHILOSOPHY TO TECHNOLOGY IS OF RELEVANCE

In the connected digital age, Stoic philosophy as a systemic unified practical philosophy of human flourishing and well-being, which combines epistemic, ethical and eudaimonic principles clustered around the central concept of wisdom, is eminently relevant and able to respond and motivate both theoretically and practically a solution to the control problem of technology.

In keeping with the Stoic principles of living virtuously, rationally, in accordance with nature, for our individual and collective well-being, as cosmopolitan citizens with a comic consciousness for the attainment of well-being, we have a collective responsibility, therefore, to respond to the problem of Big Tech's monopolistic control over the digital environment that has an overall dis-demonic impact on our individual and collective lives.

As a secular cosmopolitan philosophy, with its naturalistic and public outlook and methodology, as a way of life, Stoicism has for some time now been experiencing a renaissance both in the academy and recently in the public domain through a strong worldwide online presence. The DOIT-Wisdom model we shall examine in chapter 5, further illustrates the application of Stoic and neo-Stoic philosophy to technology. The Stoic cosmopolitan global community, therefore, has the capacity and the ability to generate and motivate a worldwide public response to the control problem of surveillance by the Big Tech companies. As a long-time fan of *Star Trek*, let me add that *Resistance is not Futile*.[4]

As my friend and colleague Rick Benitez, at the University of Sydney told me recently when the pandemic lockdown was first introduced, "Stoic Philosophy is the only game in town." I will not dispute that rhetorical flourish, though I will add that Aristotelian and Epicurean philosophy, as well as Buddhism and Hindu philosophy, as well as some indigenous philosophies, with which Stoicism has some suggestive similarities in regard to nature, are not excluded by Stoic and neo-Stoic philosophy.

NOTES

1. See also Galloway, Scott. (2017). *The Four: The Hidden DNA of Amazon, Apple, Facebook and Google*. London, Great Brittain: Bantam Press; Pasquale, Frank. (2015) *The Black Box Society: The Secret Algorithms That Control Money and Information*. Cambridge, MA: Harvard University Press; Tufekci, Zeynep. (2018). Facebook's Surveillance Machine, *The New York Times*, March 19, 2018; Zuboff, Shoshana. (2019). *The Age of Surveillance Capitalism: The Fights for A Human Future At The New Frontier of Power*. London: Profile Books Ltd; Van Dijck, Poell

and De Waal. (2018). *The Platform Society: Public Values in a Connected World*. Oxford: Oxford University Press. Webb, Amy (2019). *The Big Nine: How the Tech Titans and Their Thinking Machines Could Warp Humanity*, Hatchet Books; Lynch, Michael P. (2016). *The Internet of Us: Knowing and Understanding Less in the Age of Big Data*. Penguin Random House; Foroohar, Rana. (2019). *Don't Be Evil: The Case Against Big Tech*.UK: Allen Lane, Random House.

2. The phrase Faustian -Midas tune, alludes to the story of Faust who sold his soul to the devil for knowledge that was immortalized by Goethe in his narrative poem, Dr Faustus. The myth of Midas refers to King Midas who to satisfy greed turned everything he touched to gold. This, however, proved self-defeating as he turned his food and all those he cared for, also into gold. The allusion is meant to highlight our own part as users in using the Big Tech platforms, such as those of Facebook and Google for free, in exchange of our information, and the part of Big Tech in turning our information into gold for them.

3. For a recent book that puts forward a good case for privacy, see *Privacy Is Power*, 2020, by Carissa Véliz.

4. The phrase relates to the command issued by the *Borg* "resistance is futile," in the *Star Trek* TV series, to all those they were about to assimilate within their collective hive, by removing their individuality and all personal defining characteristics and memories, in effect rendering them willing slaves of the collective hive. The story of the Borg should also remind us of George Orwell's novel, *1984*.

REFERENCES

Annas, Julia. 1993. *The Morality of Happiness*. New York: Oxford University Press.
———. 1999. *Platonic Ethics, Old and New*. Ithaca: Cornell University Press.
Becker, Laurence. 1998. *A New Stoicism*. Princeton: Princeton University Press.
Cicero, Tusculan Disputations (Loeb Classical Library) 2nd Edition, translated by J. E. King. Cambridge, MA, and London: Harvard University Press, 1927.
Epictetus. 2018. *Epictetus: Discourses and Selected Writings*, translated and edited by Robert Dobbin. London: Penguin Classics.
Foroohar, Rana. 2019. *Don't Be Evil: The Case Against Big Tech*. London: Allen Lane, Random House.
Friedman, Madelyn. 2000. "Educating for World Citizenship," *Ethics* (April 2000) p. 589.
Galloway, Scott.2017. *The Four: The Hidden DNA of Amazon, Apple, Facebook and Google*. London: Bantam Press.
Gewirth, Alan 1998. *Self–fulfillment*. Princeton: Princeton University Press.
———. 1996. *The Community of Rights*. Chicago: University of Chicago Press.
———. 1978. *Reason and Morality*. Chicago: University of Chicago Press.
Hadot, Pierre 1995. *Philosophy as a Way of Life*. Oxford: Blackwell Publishing.
Laertius, Diogenes. 1925. *Lives of Eminent Philosophers, Book 6-10*. CM: Harvard University Press.

Long, A. A. 1986. *Hellenistic Philosophy – Stoics, Epicureans, Sceptics*. Berkeley: University of California Press, 2nd Edition.

Lynch, Michael P. 2016. *The Internet of Us: Knowing and Understanding Less in the Age of Big Data*. London: Penguin, Random House.

Nussbaum, Martha. 1997. *Cultivating Humanity: A Classical Defence of Reform in Liberal Education*. Cambridge, MA, and London: Harvard University Press.

———. 1994. *The Therapy of Desire: Theory and Practice in Hellenistic Ethics*. Princeton, NJ: Princeton University Press.

Pasquale, Frank.2015. *The Black Box Society: The Secret Algorithms That Control Money and Information*. Cambridge, MA: Harvard University Press

Plato. 1978. *The Apology*, in Edith Hamilton and Huntington Cairns (Eds.), *The Collected Dialogues of Plato*. Princeton: Princeton University Press, pp. 575–844.

Spence, Edward, H. 2011. "The Theatre of Philosophy: A Neo–Socratic Method for Performing Public Philosophy," in Bull, Gordon (Ed.), *Creativity: brain – mind – body: A view into the future of Australian Art and Design Schools, the 2011 Australian Council of University Art and Design Schools (ACUADS) Conference*, the Australian National University, the University of Canberra, and the Canberra Institute of Technology, 21–23 September 2011.

———. 2007. "Positive Rights and the Cosmopolitan Community: Right–Centred Foundations for Global Ethics." *Journal of Global Ethics*, Volume 3, Number 02, July/August 2007, 181–202.

———. 2006. *Ethics Within Reason: A Neo–Gewithian Approach*. Maryland: Lexington Books, a division of Rowman and Littlefield, 433–442.

———. 2004. *"Philosophy Plays: A Neo–Stoic Method for Teaching Ethics."* Utah: *Teaching Ethics* Volume 5, Number 1, 41–57.

Striker, Gisela. 1992. "Stoicism" in Lawrence C. Becker and Charlotte B. Becker, Editors, *Encyclopedia of Ethics*. New York: Garland Publishing, 1208–1213.

———. 1996. *Essays On Hellenistic Epistemology and Ethics*. Cambridge: Cambridge University Press.

Tufekci, Zeynep. 2018. Facebook's Surveillance Machine, *The New York Times*, March 19, 2018.

Van Dijck, Poell and De Waal.2018. *The Platform Society: Public Values in a Connected World*. Oxford: Oxford University Press.

Véliz, Carissa 2020. *Privacy is Power: Why and How You Should Take Back Control of Your Data*. London: Bantam Press, 2020.

Webb, Amy. 2019. *The Big Nine: How the Tech Titans and Their Thinking Machines Could Warp Humanity*. New York: Public Affairs, Hatchett Book Group.

Zuboff, Shoshana. 2019. *The Age of Surveillance Capitalism: The Fights for a Human Future At The New Frontier of Power*. London: Profile Books Ltd.

Chapter 5

Wisdom and Well-Being
The Dual Obligation Information-Wisdom Theory

INTRODUCTION

Where is the wisdom we have lost in knowledge?[1] This chapter[2] provides a general philosophical groundwork for the theoretical and applied normative evaluation of information generally and digital information specifically in relation to the good life for the attainment of well-being. As we saw in chapters 3 and 4, a good life conceived by the Stoics as a life of virtue in accordance with nature was both a necessary and sufficient condition for the attainment of eudaimonia or well-being. Also following on from the question posed in chapter 2, "What Is Technology Good For?" the overall aim of this chapter is to address more specifically the question of how the ethics of information and communication technologies (ICTs), as well as the ethics of AI technologies (abbreviated to t*echnology*, henceforth), can be expanded to include more centrally the issue of its *eudaimonic* impact of technology on well-being for individuals, and society. To answer that question, the chapter provides by way of a theoretical groundwork, the concept of *wisdom* understood as a type of *meta-knowledge* as well as a type of *meta-virtue*, which can enable us as individuals and collectively as a society, and humanity generally, to both know in principle what constitutes a good life, and how to successfully apply that knowledge as *know-how* in living such a life in practice, for the ultimate purpose of attaining well-being or eudaimonia. This answer will be based on the main argument presented in this chapter that the notion of *wisdom* understood as a triadic concept, at once a *meta-epistemological*, *meta-ethical* and *meta-eudemonic*, provides the essential conceptual link between information on the one hand and well-being on the other. In the digital age of information, both the theoretical examination and analysis of the question of how information in its technological construction and dissemination relates to well-being, as well as the provision of an adequate answer to that question, are essential for developing a deeper understanding of how to evaluate the theoretical and practical implications and impact of digital information on well-being for individuals and societies.

In offering this theoretical groundwork the chapter will not provide a detailed examination and evaluation of specific normative issues, which arise in the production, dissemination and use of digital information in particular contexts and instances, as that will be the topic of examination in chapters 6–9 of this book. This chapter will, however, provide in section, Wisdom and the Evaluation of Digital Information in Relation to a Good Life, a methodological approach of how different *types* of some major practical manifestations of digital information and its related AI issues can be evaluated using the meta-theoretical framework proposed in this chapter.

The chapter comprises three main inter-related parts. Part (2), A Universal Model for Evaluating the Normative Quality of Digital Information, provides a summary of an argument whose primary aim is to demonstrate a meta-philosophical model, the *Dual Obligation Information Theory* (DOIT) (Spence, 2009) to be used in the analysis and evaluation of digital information in terms of its inherent normative categories. Those categories are the *epistemological and the* ethical. That argument will, in turn, motivate and advance the main argument of this chapter, that the triadic notion of *wisdom*, at once meta-epistemic, meta-ethical and meta-eudemonic, provides the essential theoretical term that conceptually links digital information and its related AI algorithmic manifestations, to the good life and well-being.[3] This forms the overall objective of Part (3), Information, Knowledge and Wisdom, which is the examination and evaluation of the theoretical and practical relationship between *information, knowledge* and *wisdom*, based on the *Dual Obligation Information-Wisdom Theory (DOIT-Wisdom* (Spence, E. 2011).[4] The main argument presented in this chapter is that the notion of *wisdom* understood as being at once a *meta-epistemic*, *meta-ethical* and *meta-eudemonic* concept, provides the essential conceptual link between information on the one hand and the good life in the attainment of eudaimonia or well-being on the other. in the digital age of information, both the theoretical examination and analysis of the question of how information in its technological construction and dissemination relates to the good life and the attainment of well-being, as well as the provision of an adequate answer to that question, are both essential for developing a deeper understanding of how to evaluate the theoretical and practical implications and impact of information on well-being, for individuals and societies generally.

Of central importance, part (3), Information, Knowledge and Wisdom, which provides the focal direction of the chapter, offers an innovative approach in evaluating information and its relation to the good life, through the concept of wisdom. *Wisdom* understood as a type of *meta-information* or *meta-knowledge*, which comprises also essentially, epistemic, ethical and eudaimonic features, provides a direct conceptual and practical link between the concepts of information, knowledge, the good life and well-being. More generally, it provides a direct link between ICT and AI technologies and the good life and well-being. As such, the concept of wisdom allows for a direct *normative evaluation* of the impact of the dissemination of information through ICTs and AI technologies on the good life and well-being. Following on from parts (2) and (3), part (4), Wisdom and the Evaluation of Digital Information in Relation to a Good Life, provides a theoretical rationale to demonstrate the important and relevant role that wisdom plays in the specific evaluation of *digital information*. To that end, a methodological approach is used to show how some different general *types* of practical manifestations of digital information can be normatively evaluated through the application of the concept of wisdom as developed in part (3), Information, Knowledge, and Wisdom. That approach also helps establish the close conceptual connection between the DOIT-Wisdom theory to Stoic and neo-Stoic philosophy, of which wisdom is a core essential feature.

However, before proceeding, we should first address the general question of why wisdom is at all relevant to the relationship between information and a good life? Why not proceed directly to examining the role that information itself plays in the good life and well-being? Why do we need the intermediary concept of *wisdom*? The answer is fairly simple and straightforward. Wisdom, and to repeat what was said earlier, being at once a meta-epistemic, meta-ethical and meta-eudaimonic term, provides an immediate and direct conceptual link between information (an epistemic term) and the good life (an ethical-eudaimonic term). As a meta-epistemological term, wisdom as a form of *meta-knowledge* is capable of providing the necessary *reflective knowledge* and *understanding* for evaluating and applying first-order knowledge, which includes both *knowledge that* (theoretical knowledge) and *knowledge how* (practical knowledge), in making *judgements* to reach decisions that concern and impact on different aspects of a person's life (general aspects such as capacities, constraints and circumstances). The ultimate purpose of which is to assist and guide that person, and society more generally collectively, in living a good life for the attainment of well-being. For what would be the point of choosing and leading a life that was not at least in principle capable of providing a good life for the attainment of well-being? Or at least, a life that is overall better with wisdom than without it.

A basic pre-supposition of this chapter is that the presence of wisdom is, all things being equal, a better guide for a good life than its absence or its semantic and conceptual opposite, folly.[5] The notion of wisdom that informs the discussion of this chapter is developed in greater detail in part (3) is conceived as forming a continuum comprising degrees of wisdom. At the two extreme conceptual ends of the continuum, we have wisdom (good = 1) and folly (bad = 0). Most people, I speculate, probably fall somewhere in the middle – they are neither wise nor foolish.

Yet, another reason why wisdom is relevant to an enquiry concerning the role that information plays in a good life is that some of the issues with regard to the use of digital information qua good life relate not to moral or immoral conduct by an individual towards others but to the prudentially appropriate or inappropriate conduct of the individual in relation to themselves. For example, if an individual posts compromising pictures of themselves on Facebook, Instagram or YouTube, that causes no harm to anyone else but themselves, then the matter is not one of ethics, as such, but one of wisdom or at least prudence[6] – how wise or prudent was the individual to do so, if the outcome of their unwise or *foolish* behaviour in *cyberspace* turned out to be harmful to themselves in some way – say, as a result of their unwise or foolish conduct that individual was fired from a job that they enjoyed and for which they showed great potential.

In sum, the chapter will seek to show that although related, wisdom is conceptually distinct and conceptually different from both information and knowledge in a crucial way. For wisdom, unlike information and first-order knowledge, provides a person with *understanding* concerning the *techne viou* or *craftsmanship of living* in the sense of *knowing how* to *evaluate and apply* relevant information or knowledge for living a good life for the attainment of eudaimonia, and in addition, an appreciation in *knowing why* such a life constitutes a good life. You will recall that for the Stoics, and generally for the Hellenistic philosophers, philosophy was conceived as a way of life or *techne viou*. This view of wisdom, as enabling a life worth living, is also similar to the notion of wisdom defended by Sharon Ryan who in her article "What is Wisdom?" concludes that an accurate answer to that question is that "S is wise if S knows, in general, how to live well and if S has a general appreciation of the true value of living well" (Ryan 1999, 119–39). Ryan's views on wisdom will be discussed in more detail in part (3), Information, Knowledge, and Wisdom, of this chapter.

A UNIVERSAL MODEL FOR EVALUATING THE NORMATIVE QUALITY OF DIGITAL INFORMATION

The object of this part of the chapter is to describe and demonstrate in summary a meta-theoretical framework for the normative evaluation of digital information in terms of its inherent epistemological and ethical categories. The primary aim for doing so is to demonstrate in the first instance the conceptual relationship between information and knowledge and the epistemological and ethical commitments to which information understood as a type of knowledge gives rise. As aforementioned, the argument for the inherent normative structure of information as a process of communicative action has been examined and demonstrated at length in a previous publication devoted exclusively to that topic (Spence, E., 2009).[7] In this part of the chapter, the primary aim is to provide a summary of that argument sufficient for establishing an initial theoretical standpoint from which to further motivate and develop the normative relationship between information, knowledge, wisdom and the good life that forms the main topic of discussion in part (3), Information, Knowledge, and Wisdom. The demonstration of that normative relationship, at once *epistemological, ethical* and *eudemonic*,[8] will establish and demonstrate a direct normative link between information and the good life and well-being. Its purpose is to provide a methodological approach in terms of a wisdom theoretic model for the normative evaluation of that relationship and its implications and impact for individuals and societies generally.

The Inherent Normative Structure of Information and Knowledge

In describing the DOIT used for the normative evaluation of information in terms of its inherent epistemological and ethical categories, the chapter will employ an epistemological account of *semantic information* based on a minimal *nuclear* definition of information (Dretske 1999, 44–45 and 86). Following Luciano Floridi (Floridi 2005), information is defined as "well-formed meaningful data that is truthful" and following Dretske information is defined as "an objective commodity capable of yielding knowledge" and knowledge, in turn, is defined as "information caused belief."

The reference to both Floridi's and Dretske's notions of *information* in this chapter is not intended for making any critical theoretical comparisons between those two accounts of information (which lies beyond the scope of this chapter). It is rather intended for highlighting the one essential element that those accounts have in common, namely, that what is necessary for both information and knowledge is truth. For information without truth is not strictly speaking information but either *misinformation* (the unintentional dissemination of well-formed and meaningful false data) or *disinformation* (the intentional dissemination of false "information").

Using the minimal account of information described above, we can now develop an *inherent normative account of information*, which demonstrates and describes the generic epistemological and ethical commitments that necessarily arise in the dissemination of semantic information, specifically as a process of *communication*.[9]

Briefly, the argument is as follows (Spence, E. 2009): insofar as information is a type of knowledge (it must be capable of yielding knowledge, one must be able to learn from it) it must comply with the epistemological conditions of knowledge, specifically, that of truth. And insofar as the dissemination of information is based on the justified and rightful expectation among its disseminators and especially its users that such information should meet the minimal condition of truth, then the disseminators of information are committed to certain widely recognised and accepted epistemological criteria. Those epistemic criteria will in the main comprise the objectivity as well as the independence, reliability, accuracy and trustworthiness of the *sources* that generate the information. The epistemology of information, in turn, commits its disseminators to certain ethical principles and values, such as honesty, sincerity, truthfulness, trustworthiness and reliability (also epistemological values), and fairness, including justice, which requires the equal distribution of the informational goods to all citizens. Thus, in terms of its dissemination, as a process of communication, information has an intrinsic normative structure that commits everyone involved in its creation, production, search, communication, consumption and multiple other uses, to epistemological and ethical norms. These norms being intrinsic to the normative structure of information with regard to all its disseminating modes are rationally unavoidable and thus not merely optional.

Information and Universal Rights

The object of the following associated argument to the one above is to show that in addition of committing its disseminators to unavoidable epistemological and ethical standards by virtue of its own inherent normative structure in terms of truth, information as a process of *communication* commits its disseminators to respect for peoples' rights to freedom and well-being. This is by virtue of the inherent normative structure of *action* and specifically *informational action*, due to its essential features of freedom and well-being (Spence, E. 2009; 2006; and Gewirth, A. 1978).[10] Insofar as the communication of information constitutes a type of informational action, information as a process and product of communication must not be disseminated in ways that violate peoples' fundamental rights to freedom and well-being (*generic rights*), individually or collectively, or undermine their capacity for self-fulfillment.

In addition, information must as far as possible be disseminated in ways that secure and promote peoples' generic rights and capacity for self-fulfillment (positive rights) when those rights cannot be secured or promoted by the individuals themselves and can be so secured and promoted at no comparable cost to its disseminators. As we saw in chapter 3, the concept of *self-fulfillment* that I am using here, is the neo-Stoic equivalent of the Stoic concept of *eudaimonia* that I ascribe to Alan Gewirth in that chapter.

Due to constrains of space, and because it is beyond the scope and focus of this book on Stoic and neo-Stoic philosophy and its application to technology, I will not attempt to provide a detailed justification for Alan Gewirth's argument for the principle of generic consistency (PGC) on which his derivation of rights is based.[11] I will, however, offer a brief summary of the rationale of the argument for the PGC by way of a schematic outline of the three major steps of that argument.

The Rights of Agents: Alan Gewirth's Argument for the PGC

Gewirth's main thesis is that every rational agent, in virtue of engaging in action, is logically committed to accept a supreme moral principle, the PGC. The basis of his thesis is found in his doctrine that action has a normative structure, and because of this structure every rational agent, just in virtue of being an agent, is committed to certain necessary prudential and moral constraints.

Gewirth undertakes to prove his claim that every agent, *qua* agent, is committed to certain prudential and moral constraints in virtue of the normative structure of action in three main stages. First, he undertakes to show that by virtue of engaging in voluntary and purposive action, every agent makes certain implicitly evaluative judgements about the goodness of his purposes, and hence about the necessary goodness of his freedom and well-being, which are the necessary conditions for the fulfillment of his purposes. Second, he undertakes to show that by virtue of the necessary goodness which an agent attaches to his freedom and well-being, the agent implicitly claims that he has rights to these. At this stage of the argument, these rights being merely self-regarding are only prudential rights.

Third, Gewirth undertakes to show that every agent must claim these rights in virtue of the sufficient reason that he is a *prospective purposive agent* (PPA) who has purposes he wants to fulfill. Furthermore, every agent must accept that, since he has rights to his freedom and well-being for the sufficient reason that he is a PPA, he is logically committed, on pain of self-contradiction, to also accept the rational generalisation that all PPAs have rights to freedom and well-being. At this third stage of the argument, these rights being not only self-regarding but also other regarding, are now moral rights. The conclusion of Gewirth's argument for the PGC is in fact a generalised statement for the PGC, namely, that all PPAs have universal rights to their freedom and well-being (Gewirth 1978, 48–128).

Applying the PGC to information, we can now make the further argument that information generally, including both analogue and digital information, must not be disseminated in ways that violate informational agents' rights to freedom (FR) and well-being (WB), individually or collectively (negative rights). Moreover, information must as far as possible be disseminated in ways that secure and promote the informational agents' rights to FR and WB (positive rights). Conceived as the Fourth Estate, this places significant and important responsibility on the disseminators of information and in particular the media, especially journalists, both offline and online, as well as the social media.

In conclusion of this section, A Universal Model for Evaluating the Normative Quality of Digital Information, the DOIT demonstrates the dual-normative structure of *informational action,* to which all informational agents, including the media (the Legacy Media of the 4th Estate, the Digital Media of the 5th Estate (Elliott and Spence 2018) and the Big Tech Network Media of the 6th Estate such as Facebook and Google, a topic we shall examine in chapter 6, are committed by universal necessity. Information generally can be epistemologically and ethically evaluated *internally* by reference to its inherent normative structure. Insofar as the ethical values to which the inherent normative structure of information gives rise requires that the informational agents' rights to FR and WB should be respected, secured and promoted, those values are also mandated by the PGC and thus information can also be *externally* evaluated by reference to the PGC. *Expressive Information* can also be evaluated either internally or externally or both, in this way. For example, *identity theft* on the Internet is morally wrong both because it is untruthful (internal evaluation) and because it can cause harm (external evaluation).

INFORMATION, KNOWLEDGE AND WISDOM

This is the focal and main part of the chapter whose aim is the examination and evaluation of the relationship between the concepts of *information, knowledge* and *wisdom.*[12] Approximately and pending further discussion below, wisdom in this chapter is conceived as a type of *meta-knowledge* that is used in the *evaluation* and *application* of information and knowledge to make right judgements in reaching appropriate decisions that are of value and good for us in our lives personally (prudentially and eudaimonically good) and that are of value and good for others in their lives (ethically good) for the ultimate attainment of a good life resulting in eudaimonia. The overall objective of this line of enquiry is to further determine what the relationship between information and the good life is and to what degree, if any, information, in its dissemination and communication, through its related enabling ICTs and AI technologies, can in principle contribute to a good life for the attainment of eudaimonia.

In this part of the chapter, we shall examine more closely how information can be *directly* related to the notion of a good life via the concept of *wisdom*: if wisdom is a primary and essential condition for an individual in (a) determining what a good life is or ought to be (*meta-knowledge-that and meta-knowledge-why*) and (b) a primary and essential condition in providing us with guidance and direction, both as individuals and societies generally, of how to live such good lives and (c) moreover, wisdom, as a reflective meta-virtue is an enabling disposition of character that *practically enables* us to live such good lives for the attainment of eudaimonia (*meta-knowledge-how*), to what extent and in what ways, if any, does information contribute to wisdom and by extension to the good life?

The chapter posits that one direct way of evaluating the *value of information* and its relation to a good life generally (its *overall axiological goodness*) is by determining *the degree* to which it contributes or is capable of contributing to the attainment of a good life *epistemologically* (its capacity to yield knowledge), *ethically* (its ability to contribute to the moral good of others both negatively by causing no unjustified harm to others, and positively by causing positive good for others) and *eudaimonically* (its capacity to contribute to both the conception and the attainment of a good life). The chapter will show that to achieve that theoretical objective the notion of wisdom is essentially required.

Regarding the concept of a *good life*, we can say quite reasonably that a *good life* generally is one that is at least minimally capable of enabling a person to attain self-fulfillment, well-being, happiness or eudaimonia. As aforementioned, I will for methodological convenience use the term *eudaimonia*[13] to include and refer to all those concepts collectively while maintaining the original meaning for that term as intended by the ancient Greek philosophers, including Plato, Aristotle and the Hellenistic philosophers and in particular the Stoics, whose philosophy was the topic examined in chapter 3. Although those philosophers might have explained the notion and attainment of eudaimonia in different ways, they all at least agreed that the virtues were essential (and for the Stoics also sufficient) for the attainment of a eudaimonic or a flourishing life and moreover, the virtues were constitutive of such a life. For insofar as eudaimonia is the ultimate object in life as Aristotle claimed, it is difficult to conceive a life that was not at least capable of leading to the attainment of eudaimonia, as good – what would it be good for if it were incapable of at least in principle enabling one to realise one's ultimate objective in life?

Analysing information through the application of the concept of meta-knowledge (knowledge-that, knowledge-why and knowledge-how) of what is good or evil for us and others – how it contributes or is capable of contributing to a good life for us and others for the attainment of eudaimonia – is what the chapter will initially postulate as *Wisdom*. In sum, if a good life should at least be capable of leading to self-fulfillment or eudaimonia (otherwise what is it good for?), especially, self-fulfillment as capacity fulfilment (making the best of oneself as a human being[14] then *wisdom* (understood as a type of *meta-knowledge*, the acquisition of which enables one to create, communicate and use information so as to render oneself and others, whenever possible, capable of achieving self-fulfillment and eudaimonia) is a *necessary condition for a good life*.

An important qualification to the claim made in this chapter that wisdom is a necessary condition for a good life is that such a life is conceived eudaimonically. For the notion of wisdom developed in this chapter and applied in evaluating the *axiological goodness* of information is itself a eudaimonic conception of wisdom. However, such a eudaimonic notion of wisdom is not unlike our common sense and pre-theoretical understanding of wisdom, namely, an overarching reflective capacity the possession of which allows one to lead a good life and, moreover, enables one to guide others in leading fulfilling and good lives. This eudaimonic notion of wisdom is akin to the notions of wisdom defended by philosophers such as Plato, Aristotle, the Epicureans, and as we saw in chapter 3, by the Stoics and neo-Stoics, who although postulated and defended somewhat different notions of the good life, can nevertheless collectively be thought of as offering eudaimonic accounts of the good life.

It can be said that a common denominator for all the cited eudaimonic accounts of wisdom is their subordination of the concepts of pleasure and desire to that of virtue. The essential link between pleasure and desire on the one hand and virtue on the other might be weaker in the case of the Epicureans and stronger in the case of the Stoics but whatever the strength of that relationship might be, the link between pleasure, desire and virtue is an essential characteristic of traditional eudaimonic conceptions of a good life and also eudaimonic conceptions of wisdom. Moreover, as mentioned above, another essential characteristic of a eudaimonic conception of a good life is that such a life is capable of resulting in the attainment of eudaimonia. As such, the term *eudaimonic* understood in this very general sense is compatible with a number of other theories of the good life which can be shown to be capable of leading to the attainment of eudaimonia. This is in keeping with the *Eudaimonist axiom*, the view that "happiness is desired by all human beings as the ultimate end or telos of all rational action."[15]

Importantly, the relationship between wisdom and a good life proposed in this chapter under a eudaimonic conception of a good life is *reflexive*. For wisdom guides one to the choice of a eudaimonic conception of a good life and the pursuit of such a life, and a eudaimonic conception of a good life, in turn, guides and motivates one to the acquisition of wisdom as an *enabling disposition*, in the form of an overarching reflective virtue, which is necessary for the attainment of a eudaimonic life. This should not surprise us. For although wisdom acts initially instrumentally, as a necessary enabling virtuous disposition for the attainment of a eudaimonic life, once attained, a eudaimonic life becomes inseparable from the state of wisdom that enabled its attainment. This reflexivity between eudaimonia and wisdom allows us then to say that a wise person is generally a eudaimonic person and a eudaimonic person is generally a wise person. However, I don't wish to exclude the logical possibility that one could be wise but unhappy although pragmatically, given our common understanding of wisdom, that would be an odd thing to say and in practice I think, an unusual occurrence. For as we saw in chapter 3, the Stoics claimed that the virtues, including practical wisdom or prudence, was not only necessary for a eudaimonic life but also sufficient. As we also saw in chapter 3, the neo-Stoic philosophy I am attributing to Alan Gewirth through his theory of self-fulfillment, also supports the sufficiency of virtue for a eudaimonic life.

What Is Wisdom?

Having examined in some detail what information and knowledge are and what the relationship is that exists between them (by way of an examination of the essential property that characterises both, namely, the property of truth) it is now time to turn our attention to the notion of wisdom so as to explore further the conceptual relationship that holds between information, wisdom and a good life.

According to Nicholas Maxwell (2007, 79),

> The central task of inquiry is to devote *reason* to the enhancement of *wisdom* – wisdom being understood here as the desire, the active endeavour, and the capacity to discover and achieve what is desirable and of value in life, both for oneself and for others. Wisdom includes knowledge and understanding but goes beyond them in also including: the desire and active striving for what is of value, the ability to see what is of value, actually and potentially, in the circumstances of life, the ability to experience value, the capacity to help solve those problems of living that arise in connection with attempts to realize what is of value, the capacity to use and develop knowledge, technology and understanding as needed for the realization of value. Wisdom, like knowledge, can be conceived of, not only in personal terms, but also in institutional or social terms. We can thus interpret the philosophy of wisdom as asserting: the basic task of rational inquiry is to help us develop wiser ways of living, wiser institutions, customs and social relations, a wiser world.

What is of interest in Maxwell's quoted passage for our present purposes is the relationship he draws between the concepts of *reason*, *knowledge*, *understanding* and the *desire*, *capacity*, and *active endeavour* for the *achievement* (or as in my case *attainment*) *of* what is *of value in life, for oneself and others*. With the exception of *understanding*, for which I will have more to say in what follows, the other concepts to which Maxwell draws attention seems to anticipate and reflect both explicitly and implicitly, the concepts included in my own normative analysis of information and knowledge, in terms of their epistemological, ethical and eudaimonic dimensions. The basis of that analysis is the meta-theoretical framework comprising the DOIT and as we shall see in section 4, The DOIT-Wisdom model.

Four Theories of Wisdom

In an article in the *Stanford Encyclopedia of Philosophy*, Sharon Ryan (2007) identifies at least four different theories of wisdom: (A) *Wisdom as Epistemic Humility*, which she attributes to Socrates in Plato's *Apology* (20e-23c). There Socrates expresses puzzlement concerning the oracle of Delphi's pronouncement that he is the "wisest of men" and declares that his knowledge extends only as far as his knowledge of his own ignorance, the prototype case of *epistemic humility* (B) *Wisdom as Epistemic Accuracy*, for which Ryan provides two versions to the effect that (B1) S is wise iff for all p, (S believes S knows p iff S knows p) (EA1) or the weaker version (B2) S is wise iff for all p, (S believes S knows p iff S's belief in p is highly justified.) (EA2); (C) *Wisdom as Knowledge*: the view that knowledge is at least a necessary condition of wisdom. Ryan identifies several philosophers who hold some version of (C) including Aristotle (*Nichomachean Ethics* VI, Ch. 7), Descartes (*Principles of Philosophy*), Richard Garrett (1996), John Kekes (1983), Lehrer and Smith (1996), Nicholas Maxwell (1984), Robert Nozick (1989), Plato (*The Republic*) and Sharon Ryan (1996,1999). According to Ryan, all these philosophers "have theories of wisdom that require a wise person to have some knowledge of some sort" and what's more, "all these views maintain that wise people know what is important" (Ryan, 2007). Overall, these theories differ according to Ryan "over what it is that the wise person must know and whether there is any action that is required for wisdom" (Ryan, 2007).

To further differentiate different notions of wisdom that fall under the broad category of "wisdom as knowledge," Ryan refers to Aristotle who held two main theories of wisdom, **Sophia**, or *Theoretical Wisdom*, and **Phronesis** or *Practical Wisdom*. Theoretical wisdom according to Aristotle is "scientific knowledge, combined with intuitive reason, of the things that are highest by nature" (Ryan 2007 quoting Aristotle in *Nichomachean Ethics*, VI, 1141b). On that basis, Ryan interprets Aristotle's notion of theoretical wisdom as the following view: (C1) *Wisdom as extensive factual knowledge (WFK)* which in effect amounts to "S is wise iff S has extensive factual knowledge about science, history, philosophy, literature, music, etc" (Ryan, 2007). Ryan finds this notion of wisdom implausible for as she correctly observes "some of the most knowledgeable people are not wise" (Ryan, 2007). Ryan maintains that Aristotle's notion of *phronesis* or practical wisdom is a more reasonable theory to hold. According to Aristotle,

> Now it is thought to be the mark of a man of practical wisdom to be able to deliberate well about what is good and expedient for himself, not in some particular respect, e.g. about what sorts of things conduce to health or strength, but about what sorts of things conduce to the good life in general. (*Nichomachean Ethics*, VI, 1140a01140b)

Ryan concludes that "for Aristotle, practical wisdom requires knowing, in general, how to live well." She goes on to say that although many philosophers are in agreement with Aristotle on this point they would not agree with Aristotle "that theoretical wisdom is one kind of wisdom and practical wisdom another. Wisdom, in general," Ryan concludes, "requires practical wisdom" (2007). In support of this general notion of practical wisdom, Ryan cites Robert Nozick who claims that "Wisdom is what you need to understand in order to live well and cope with the central problems and avoid the dangers in the predicaments human beings find themselves in" (Nozick 1989, 267). She also cites John Kekes, whose view is that "what a wise man knows, therefore, is how to construct a pattern that, given the human situation, is likely to lead to a good life" (Kekes 1983, 280). Ryan defines this type of general practical wisdom as (C2) *Wisdom as Knowing How to Live Well (KLW)*: "S is wise iff S knows how to live well," which according to Ryan captures on the whole Aristotle's concept of practical wisdom as well as the views held by Nozick, Plato, Garrett, Kekes and Ryan (2007).

The final theory of wisdom that Ryan considers is *Wisdom as Knowledge and Action*, which she specifically defines as *Wisdom as Knowing How To, and Succeeding at, Living Well (KLS),* which in effect amounts to: S is wise iff (i) S knows how to live well, and (ii) S is successful at living well (2007). According to Ryan, the "idea of the success condition [condition (ii) in KLS] is that one puts one's knowledge into practice." She goes on to attribute a view broadly along the lines of (KLS) to Aristotle (his notion of practical wisdom), as well as to Kekes and Nozick. Ryan herself rejects this theory based on criticisms she raises in (Ryan 1999) but her main criticism with which I concur is that (KLS) seems to leave out the factual knowledge required by the theory of WFK.

In agreement with Ryan, I also claim that some factual knowledge of the world (but not necessarily extensive) adequate for enabling a person to make their way in the world and have a good life, is necessary for wisdom. This consideration introduces an important distinction when enquiring into the conceptual connection between knowledge and wisdom: the distinction between *knowledge for wisdom* and *knowledge as wisdom*. Although related, the two are quite different and their difference highlights an important and crucial distinction between knowledge and wisdom.

No doubt some general knowledge about the world acquired on the basis of reliable and veridical information that causes it and sustains it (Dretske 1999) is necessary for wisdom. This is in keeping with the notion of Socratic ignorance, roughly understood here as having knowledge of one's ignorance (being aware of one's ignorance and humbly acknowledging one's lack of knowledge). For Socratic ignorance prompts and motivates one to acquire the knowledge of which one is ignorant (knowledge understood here as some minimal general knowledge about some basic aspects of the world, e.g., history, geography, science, mathematics, literature, art, etc.). Socratic ignorance as a special type of knowledge accords with Ryan's theory of wisdom as epistemic humility, which was examined earlier. By contrast, those who claim to know what they lack knowledge of, are not in a position to be motivated to acquire the knowledge they lack; and moreover, the knowledge that is at least in a minimal and general sense partly necessary for the acquisition of wisdom and by extension, the attainment of a good life and eudaimonia.

Thus, at a minimum, and bracketing the possibility that "holy fools" though totally ignorant of facts about the world are nevertheless in some sense "wise," some minimal and general knowledge about the world is instrumentally and prudentially necessary for the acquisition of wisdom. At least at a minimum, an attitude of Socratic ignorance[16] might be necessary for the acquisition of wisdom. For the Socratic elenchus can be applied as a method for acquiring the knowledge one lacks, through first recognising and acknowledging one's ignorance, and then being motivated to gradually acquire the knowledge of which one is ignorant, through critical enquiry and further investigation. According to John Kekes, "the elenchus enables its practitioners to progress from a special kind of ignorance – foolishness – to a special kind of knowledge – moral wisdom" (Kekes1995, 39).

We can therefore say that the acquisition of such general minimal knowledge about the world or an attitude of Socratic ignorance when we lack such knowledge is instrumental to the acquisition of wisdom because it provides at least part of the necessary means, that is, the capacity for the acquisition of wisdom. Moreover, the acquisition of such minimal and general knowledge of the world or in its absence, an adoption of an attitude of Socratic ignorance, is *prudential* to the acquisition of wisdom. Insofar as we consider the acquisition of wisdom desirable, valuable and essential for the attainment of a good life, we *should* (normatively) inculcate in ourselves the virtue of learning: the desire and active pursuit of the acquisition of at least a minimal and general knowledge about the world. Hence, some minimal and general knowledge about the world is necessary *for* wisdom. The Stoics also claimed that some general knowledge of the world and more generally knowledge of nature was also required for wisdom. In the *Meditations*, for example, Marcus Aurelius speaks "of an understanding which comprehends the inmost being of each thing, its place in the world-order, the term if its natural existence, the structure of its composition" (Aurelius1964, Book 10:9, 155).

In addition, along with Aristotle, I also wish to claim that being a good person is a necessary condition, for being wise. For Aristotle claims, "Therefore it is evident that it is impossible to be practically wise without being good" (*Nichomachean Ethics*, VI 1144a). Contrary to Aristotle, however, and in accordance with the Stoics and my Gewirthian reconstruction of a neo-Stoic position I argued for in chapter 3, I wish to claim that being a good person, that is, a person in possession of a virtuous character, is also sufficient for being wise, in successfully applying wisdom in living well, at least for the aim, if not the actual successful attainment of eudaimonia, a prize reserved only for the Stoic sage, for example, someone like Socrates. John Kekes seems to also have the view that a virtuous character is an essential characteristic of the wise person. In the opening sentence of his book *Moral Wisdom and Good Lives* (1995, ix), Kekes tells us that "moral wisdom is a virtue – the virtue of reflection." A more detailed characterisation of moral wisdom by Kekes (1995, 5–7) is that,

> Moral wisdom is the capacity [a psychological capacity] to judge rightly what should be done in particular situations to make life better. . . . Because this human psychological capacity, once developed, is likely to be lasting and important, it can be identified as a character trait . . . . We can say, therefore, that people have moral wisdom if they regularly and predictably act wisely in the appropriate situations and if so acting is an enduring pattern in their lives . . . . Whether an action is morally wise depends also on what the agents bring to the judgements they make, such as their particular conception of what would make life better.

According to Kekes, moral wisdom is a second order virtue whose primary concern,

> [Is] the *development of our character* [emphasis added] in a desirable direction by strengthening or weakening some of our dispositions. First-order virtues guide our actions in view of what we think of a good life; second-order virtues guide our actions with a view of developing the kind of character that reflects a reasonable conception of a good life. (1995, 9)

Kekes' claims to moral wisdom above are also in keeping with the importance the Stoics and neo-Stoics attribute to *character* as an essential aspect of being a good and wise person.

The Psychology of Wisdom: The Berlin Wisdom Paradigm

Finally, it is worth mentioning a *psychological theory of wisdom* that runs parallel to the philosophical theories of wisdom discussed above. This is the theory of wisdom postulated by Paul Baltes and his research associates from the Max Planck Institute for Human Development in Berlin. The theory known as the *Berlin Wisdom Paradigm* views wisdom as a kind of *expertise* in the matters of human life (Baltes 2004, Baltes et al. 2002, Baltes and Smith 1990 as referred to by Banicki 2009). In a chapter from *A Handbook of Wisdom: Psychological Perspectives* (Robert J. Sternberg and Jennifer Jordan 2005, 115), Ute Kunzmann and Paul B. Baltes define wisdom in accordance with the Berlin Wisdom Paradigm, as a "highly valued and outstanding expertise in dealing with fundamental, that is, existential, problems related to the meaning and conduct of life" (2005, 117).

Kunzmann and Baltes go on to elaborate that the "focus of their theoretical work has been to define wisdom as an expert system in human thought and behavior that coordinates *knowledge and virtue, mind and character*" [emphasis added] (2005, 128), and that thus defined, wisdom for them "reflects both components of wisdom: intellect and character" (2005, 130). According to them, "wisdom differs from other human strengths in that it involves an orchestration of mind and virtue, intellect and character" (2005, 131).

It is not too much of a jump to reasonably interpret the psychological notion of the Berlin Wisdom Paradigm as expressed above by Kunzmann and Baltes as parallel and broadly in keeping with the general view of practical wisdom, as knowledge of how to live well and the successful application of that knowledge in living well. However, with the additional proviso that the knowledge in question must also include some knowledge of facts concerning the world, which is Ryan's notion of wisdom (KLS) we examined above. Such knowledge does involve an "orchestration," as in the case of Kunzmann and Baltes' psychological notion of wisdom, of "intellect" or "cognition" (the epistemological features) and virtue and character (the ethical and eudaimonic features) in my tripartite model of wisdom, conceived as meta-knowledge and meta-virtue, which, in general, is in keeping with the notion of *intelligent virtue* developed by Julia Annas in her book by that title (2011). Moreover, the theories of wisdom we examined above, as well as that of the Berlin Wisdom Paradigm, are generally in keeping with Stoic and neo-Stoic philosophy we examined in chapter 3.

WISDOM AND THE EVALUATION OF DIGITAL INFORMATION IN RELATION TO A GOOD LIFE

The outcome of the extended argument concerning the nature of wisdom, what wisdom is, and its relationship to information and knowledge in part

(3), Information, Knowledge, and Wisdom, is that wisdom is a special type of meta-knowledge and meta-virtue. Insofar as the ultimate purpose of a good life is the attainment of eudaimonia then wisdom, which informs the conception of a good life and directs its active pursuit for the attainment of eudaimonia, is an essential condition for both the conception and the attainment of a good life. As the essential condition for both the conception and guided active pursuit and successful achievement of the good life, wisdom is therefore established as the essential conceptual connection between information and the good life and in particular information that is minimally necessary for acquiring knowledge of the things we need, to have a good life. Some general knowledge about the world acquired on the basis of reliable and veridical information that causes it and sustains it (Dretske 1999), is therefore necessary for wisdom. This, in turn, allows us to determine some of the generic implications and ramifications of information for the conception of a good life, in particular, a eudaimonic conception of a good life. However, as Kekes points out, "the eudaimonic conception of a good life is not to be understood as the endorsement of a particular form of life. It is rather a *regulative ideal that specifies some general conditions to which all good lives must conform*" [emphasis added] (Kekes 1995, 24). As such, the eudaimonic account of a good life canvassed in this chapter is broadly speaking pluralistic as it is in principle compatible with other different conceptions of a good life that meet the same necessary general conditions to which any notion of a good life must conform. For example, insofar as hedonistic, desire-satisfaction and objective list theories of the good life meet the minimal conditions for both specifying what a good life is, as well as providing the enabling conditions for its practical realisation, then they too can be aligned broadly to the notion of wisdom developed in this chapter.

Kekes' claim cited above is insightful and very much in keeping with the eudaimonic conception of a good life proposed in this chapter on the basis of the DOIT-Wisdom. For the *DOIT-Wisdom model* (2011) is intended only as a meta-theoretical *regulative ideal that specify some general conditions to which all good lives (and in particular informational lives) must conform* regardless of the particular contexts and contingencies of those lives.

Specifically, with regard to the creation and dissemination of information, central to those general conditions to which all good lives are bound are (a) the epistemological and ethical obligations that emanate directly from the inherent normative structure of information; (b) the universal rights to freedom and well-being to which all agents are entitled and which arise naturally from the inherent normative structure of informational action and (c), the virtues of character and associated moral sentiments and values that are prudentially desirable and required as *enabling general motivational dispositions* for the pursuit of a good life for the ultimate attainment of eudaimonia.

In sum, we can say that those general meta-conditions are encapsulated within the combined models put forth in this chapter, namely, those of DOIT and Wisdom to which I refer to collectively in this chapter as the *DOIT-Wisdom* model (Spence, 2011). That model moreover seems adequate for the normative evaluation of an *informationally good life* for the attainment of well-being, as it combines a tripartite structure comprising the three essential normative principles for the evaluation of digital information, namely, the epistemic, the ethical, including the virtues as enabling dispositions, and the eudaimonic. Kekes correctly claims that according to a eudaimonistic conception of a good life,

> Primary values [values that concern uniform and universal human goods and needs] may be thought of as establishing the moral limits and secondary values [values that vary across individuals in accordance with differences in cultural traditions, conceptions of a good life, and individual contingencies and circumstances] as establishing the moral possibilities that define good lives (1995, 25) ... the former define *a grid* [emphasis added] within which human beings must endeavour to make a good life for ourselves, while the latter provide the ways in which individuals fill in the grid. (1995, 23)

It has been the overall objective of this chapter to provide such a conceptual *grid*; namely, the DOIT-Wisdom model. Having discussed meta-theoretically the epistemological, ethical, and eudaimonic implications of information as a process of communication for individuals and society generally through the notion of wisdom in part (3), Information, Knowledge and Wisdom, I will in this final part of the chapter apply the DOIT-Wisdom model to identify some of the specific implications and ramifications that the production, dissemination and communication of *digital information* might have for the good lives of individuals and society generally. Before doing so, it would be useful to provide a schematic representation of the triadic structure of the DOIT-Wisdom model:

The DOIT-Wisdom model (Spence, 2011, 2009) enables the normative evaluation of technology in relation to well-being based on the concept of *wisdom*. The argument presented in this chapter demonstrates that the notion of *wisdom* understood as triadic notion comprising at once a *meta-epistemic*, *meta-ethical* and a *meta-eudaimonic* normative categories, as figure 5.1 illustrates, provides the essential conceptual link between technology (specifically, information and communication technologies and AI technologies) and well-being.

WHY WISDOM IS OF RELEVANCE TO THE EVALUATION OF DIGITAL INFORMATION

The model DOIT-Wisdom presented in this chapter for the normative evaluation of information and its relation to the good life is a model that

Wisdom and Well-Being 91

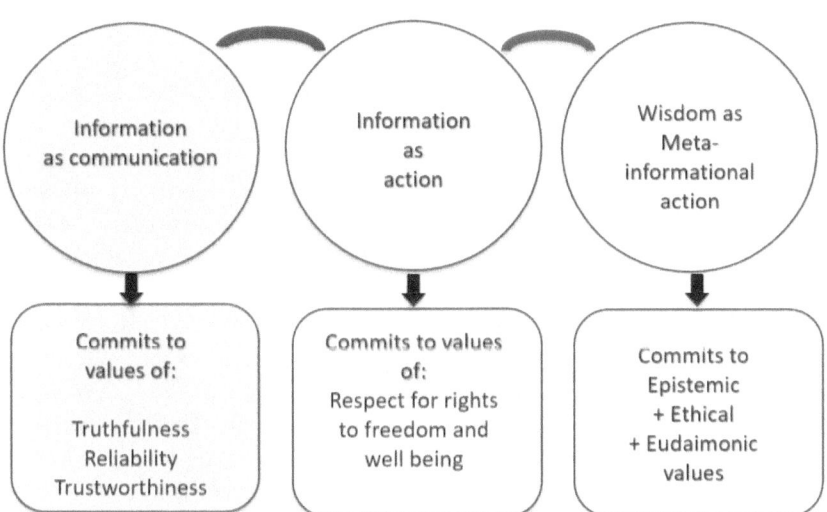

Figure 5.1 The DOIT-Wisdom Model.

applies to all information whether digital or analogue regardless of whether such information is generated within the legacy media or the digital and social media. However, the exponential growth of digital information on the Internet, through technology networks, such as those Facebook, Google, Microsoft, Apple and Amazon, collectively referred to in this book as BigTech or the Big Five, and its implications for the good life of individuals and society generally, merits special attention for at least three theoretical reasons.

There is, to begin with, a *metaphysical or ontological reason*: the digitalisation of information has extended the natural and technological boundaries of our lives to such an extent that as Luciano Floridi correctly observes we now live in the infosphere (2002) and not just the biosphere. We have essentially become and are increasingly becoming informational beings. As such, billions of people around the world spend a great deal of their lives, thinking, communicating, working, buying, selling, banking, playing, socialising and generally, *acting informationally*. All these numerous and varied informational activities in the infosphere, and specifically the digital infosphere, are unavoidably related to the quality and goodness of our lives, both as individuals and collectively as societies. Hence, if wisdom is the conceptual link that normatively links information to a good life and well-being, as argued in this chapter, clearly then wisdom has a direct and primary relevant role to play in the normative evaluation of digital information and its relationship to a good life and well-being in the infosphere. As shown in chapter 4, wisdom as a core principle of Stoic philosophy applies to the biosphere and analogously it applies equally to the infosphere.

Second, there is a *technological reason*. The digitisation of information allows us countless means and ends for instantly creating and disseminating information around the globe through the Internet and through technological devices such as smart phones, for example. This unprecedented rapid and extensive dissemination of digital information can and does affect the quality of our lives both positively but also negatively. People can both benefit themselves and others through the use of digital information, but they can also harm themselves and others through the *unreflective* misuse and abuse of information, though, for example, fake news, deep fakes and other forms of misinformation and disinformation.

The age of abundant information is paradoxically marked by a deficit of wisdom. It seems that, the more information we have, the less wise we are in managing and controlling it for our individual and collective well-being. The problem is that there is too much information and not enough time to absorb it, understand its implications and judge the best way to use it for our individual and common good. The glut of information has created gluttony for information, which can lead us to behave unwisely and sometimes foolishly. Examples of such unwise and foolish online behaviour abound. Take for example the Australian treble Olympic gold medal winner Stephanie Rise,[17] who lost a lucrative sponsorship with Jaguar as a result of a thoughtless tweet about the South African rugby team; the Canberra Raiders star Joel Monaghan[18] who was photographed performing an act of simulated bestiality with a dog, which was later published on the Internet and forced his resignation; Catherine Daveny,[19] a journalist with the *Age* who was fired for a series of unsavoury but mostly silly tweets about various TV personalities and celebrities; a journalist photographer from Ireland who was sacked after making comments on her Facebook page about the young woman Michaela Harte, who had been murdered on her honeymoon in Mauritius; American journalist Nir Rosen,[20] who had to resign after making offensive tweets about Lara Logan, the CBS journalist who was sexually attacked in Egypt. More commonly still, young people undermine their privacy and chances of future employment by placing compromising photos of themselves and friends on Facebook and engaging in cyber-bullying that has allegedly driven some young people to suicide. The problem is widespread and global.

There are countless daily informational acts of unwise and self-defeating behaviour, which, but for the all-seeing eye of the omnipresent Internet, would pass unnoticed as matters of no consequence. An important truth the WikiLeaks controversy has revealed is that the Internet and the digital informational environment affords no one the certainty or comfort of privacy. The Internet is, by its very nature, a boundary-free public space not well-suited to private conversations and secrets and, what's more to the point, informational indiscretions. The digitisation of information has fundamentally changed not only the way we disseminate information but the way we live. The Internet, just like the natural environment, cannot be neatly constrained and controlled by any one group of individuals, conglomerates or nations. And like the natural environment, the Internet is capricious and unpredictable and only favours the uninhibited free dissemination of information by anyone, anytime and anywhere.

If we cannot control or manage the flow of information on the Internet, the next best thing is to control our own online informational behaviour. That is within our control and in keeping with the Stoic and neo-Stoic philosophy examined in chapters 3 and 4, a central principle of which is wisdom. We have to learn how to use and disseminate information wisely, in a manner that protects and promotes our individual and collective well-being. Wisdom, as a *higher type of knowledge* (knowing how to understand and use information upon reflection and with good judgement for the benefit of oneself and society), provides practical "know-how" for applying information to improve our lives and that of others. It is also a *reflective virtue* in the form of practical prudence, which can teach us how to create and use information to live good and meaningful lives in the infosphere – lives that are capable of leading to self-fulfillment, eudemonia and happiness for us and others. What wisdom requires is that we learn the husbandry of information. How to reflect upon it, how to understand it, how to control it so it does not control us, how to judge its implications so we can foresee its consequences, whether they are good or bad, and how to use it in ways that enhance our well-being and promote and protect our rights to freedom, privacy and respect. In the age of information, it seems that we would be better off with more wisdom and a little less information. Switching over our iPhone and iPad to Plato and Epictetus, sometimes, may be a good start.[21]

The uncritical and sometimes thoughtless dissemination of a lot of trivial and personal information of oneself and others on social networks and platforms such as Facebook, YouTube and Instagram, as well as through SMS and Twitter, as the above examples illustrate, might be not only not conducive to wisdom and a good life overall, but detrimental to it. For they appear to encourage, sometimes at least, unreflective, foolish and reckless behaviour with no apparent compensating axiological or eudemonic value for oneself, one's "friends" and others. Given the pitfalls of creating and accessing unreflectively information about oneself and others, without any compensating realisation of value in relation to one's conception of a good life, a wise person would thus be best served in exercising caution when using digital information. Such overall reflective caution as we saw in chapter 4, is also in keeping with Stoic philosophy's essential characteristic, namely, control; that is to say, the ability to narrow down the areas in one's life over which one has little or no control so as to enlarge the areas of our lives over which we have control so that we can order them in conformity with our conception of a good life for the attainment of eudaimonia or well-being. This is a process of self-knowledge, a process of reflection involving judgements "whose aim is to make our character less fortuitous and more deliberate" (Kekes 1995, 127–28). When we allow ourselves to unreflectively misuse digital information of ourselves and others on the Internet and through the use of smart phones and other digital devices, we make ourselves more "fortuitous" and less "deliberate." In so doing, we become hostages to fortune, since we no longer have any control over that information, which now for ever floats beyond our control in cyberspace. Surely such unreflective conduct is not only not wise but can in extreme circumstances be foolish and ultimately self-defeating if, upon careful reflection, it undermines our considered conception of a good life and that of its eudaimonic realisation.

A third reason why wisdom is of primary relevance to an enquiry concerning the role that digital information plays in a good life is that some of the issues with regard to the use of digital information qua good life relate not to moral or immoral conduct by individuals towards others as such, but to the prudentially appropriate or inappropriate conduct of the individuals in relation to themselves, or more precisely what is good for them – good for their lives and well-being. In short, such conduct might only constitute unwise or foolish behaviour and not necessarily unethical conduct. For example, if an individual posts compromising pictures of themselves on Facebook that causes no harm to anyone else but themselves, then the matter is not one of ethics as such, but one of wisdom or at least prudence[22] – how wise or prudent was the individual to do so, if the outcome of their unwise behaviour in public *cyberspace* turned out to be harmful to themselves in some unintended way. Say, as a result of their unwise but not necessarily unethical conduct that individual was fired from a job they enjoyed and relied upon for their self-fulfillment.

Digital media, including our multiple informational interactions on the Internet and through technology networks, such as Facebook, Instagram, Google, YouTube and Twitter, have made it possible for us to access and use extraordinarily and unprecedented large quantities and varieties of information and data. This is indeed the age of information and Big Data. However, what seems to follow from our characterisation of wisdom above is that the uncritical access and use of so much information without the appropriate reflection, judgement and understanding, might not be conducive to wisdom and consequently might not be conducive to a good life for individuals and society. Thus, the uncritical accumulation and use of more information is not necessarily conducive to more wisdom (because not essential to it) and hence not more conducive to a good life. On the contrary, sometimes we might be better off with less information rather than more, especially if the former is directed by wisdom (less information) and the latter is not (more information).

Even in the case of critically accessing and using *contextualised information*, such information might also not be conducive to wisdom and to a good life, if that information is merely used instrumentally without a clear understanding of the ends which that information is intended to achieve or a clear understanding of the value of those ends. Recall that wisdom not only directs the means but also the ends of our actions. Thus, the accessing and use of a lot of contextualised information on our smartphones, for example, without a clear understanding of the value of the ends for which that information is to be used, might at best be neutral with regard to the goodness of our lives and worse detrimental to the goodness of our lives if it utilises too much of our cognitive and social resources for the acquisition of information that ultimately is of little or no axiological value (epistemic, ethical, or eudaimonic) for us personally and for society at large. Worse still, it can result, as we shall see in the chapters that follow (chapters 6–9), in the targeting and undermining of our autonomy, privacy, integrity and our digital identity by the Big Tech companies such as Facebook and Google, through their opaque and unaccountable algorithms that systemically target our information as users and data subjects, for their own self-seeking financial gain.

CONCLUSION

The object of this chapter has been to explore how information generally and digital information specifically can be normatively evaluated in relation to the concept of a good life and well-being. To that end, the chapter developed and applied a meta-theoretical model, the DOIT-Wisdom model, as a groundwork for demonstrating how information can be normatively evaluated, using a triadic cluster of inter-related normative categories, those of the epistemic, ethical, and eudaimonic categories, including the virtues that are integrally related to those axiological categories, analysed, and defended on the basis of that model and on the basis of Stoic and neo-Stoic philosophy in chapters 3 and 4. Central to those normative categories is the concept of wisdom, which being at once a meta-epistemological, meta-ethical and meta-eudaimonic concept, establishes a direct conceptual link between information and the notion of a good life and well-being both for individuals and societies generally. More specifically, wisdom as this chapter has demonstrated has a particular and important normative relevant role to play in the normative evaluation of digital information. The cited examples of the dissemination of digital information on the Internet, including other digital media and networks in part (4), Wisdom and the Evaluation of Digital Information in Relation to a Good Life, were used illustratively as proof of concept to demonstrate how the notion of wisdom as the essential conceptual link between information, a good life and well-being can be used *methodologically* to evaluate axiologically (epistemically, ethically an eudaimonically) the creation, access, use and dissemination of any type of digital information on the Internet. Some digital informational evaluations such as, for example, the use of *Skype* or Zoom to communicate regularly with close friends, family and colleagues, will be positive, given that relationships with friends, family and colleagues are conducive to a good life and well-being and as such, their cultivation and maintenance, especially during COVID-19 pandemic, is generally, *wise-worthy*. Other digital informational evaluations, such as the examples from Facebook, SMS and Twitter above, may prove negative because not conducive to a good life and well-being and possibly conducive to the diminution of well-being, and therefore *not wise-worthy*.

Although *descriptively* it can be said that we are now living in the age of information, we should act in a *protreptic* manner if we are wise or at least aspire to be wise, so as to promote the *Age of Wisdom*, both for our own eudaimonic good and that of others, and especially for the sake of future generations who might mistake mere information and big data for knowledge and mere knowledge for wisdom. This chapter has argued for a theoretical groundwork for the normative evaluation of information generally, and digital information specifically through the notion of wisdom, a key principle of Stoic philosophy and generally Greco-Roman and Hellenistic philosophy. No doubt a lot more work must be accomplished in a more extensive examination and evaluation of specific types of good and bad uses of information that daily impact on the lives of people both positively and negatively. That is a topic we will examine in chapters 7–9 of this book. As we adapt ourselves to live more and more in the infosphere, including virtual worlds, this becomes a desirable and indeed a necessary task. I hope this is but the beginning of an ongoing research project.[23]

NOTES

1. This line is from a poem from T. S. Eliot, Choruses from The Rock, 1934.

2. This chapter has its conceptual origins in a three-year research project I undertook in the Department of Philosophy at the University of Twente, Netherlands, 2006–2009, as a Research Fellow of a VICI International Research Project for the *Evaluation of New Media in Relation to the Good Life*. The project was funded by the National Dutch Research Council (NWO) under the supervision of Professor Philip Brey.

3. Henceforth, understand the use of the concept of *information*, used in this chapter, to relate to digital information generally, including its related manifestations of production and dissemination via ICTs and AI technologies, including big data and algorithms.

4. This chapter is based on an earlier version of the paper published by Spence E. 2011: Information, Knowledge and Wisdom: Groundwork for the Normative Evaluation of Digital Information and Its Relation to the Good Life, *Ethics and Information Technology*, (2011) 13:261–275. DOI 10.1007/s10676-011-9265-7.

5. Erasmus of course in *Praise of Folly* makes a case, albeit an ironic and satiric one, that Folly is in fact a better guide than Wisdom for having a good life *in this world*. On the serious side, Erasmus, however, goes on to suggest in the last section of *In Praise of Folly* that since only God is capable of wisdom and human beings only capable of folly we should submit ourselves to the will and guidance of God. Being primarily a secular examination of the role of wisdom in a good life, this chapter will not, however, pursue that theological line of inquiry, interesting though it might be. However, the role the Stoic notion of nature plays in a good life for the attainment of eudaimonia, as examined previously in chapters 3 and 4, is suggestive and of relevance in the neo-Stoic version of that concept presented in those chapters.

6. The term *prudence* used throughout this chapter refers to the virtue of prudence conceived as an enabling disposition or trait of character that has the tendency of preventing an individual of engaging in conduct that is likely to cause them harm. That is, conduct unbecoming of a virtuous person. Unlike the term *instrumental rationality*, which is a non-moral term and refers to the non-moral self-interest of an individual, prudence by contrast refers to the virtuous and hence moral self-interest of an individual person.

7. The paper Spence E (2009) provides the basis for this part of the chapter.

8. These three normative terms, the epistemological, ethical and eudaimonic, will sometimes be referred to in this chapter by the collective term, *axiological*, to be understood as, *normatively conferring values*.

9. For a more detailed account of the epistemological and ethical commitments to which information as a process and product of communication gives rise, see Spence, E. (2009).

10. Alan Gewirth's main thesis in *Reason and Morality* (1978) is that every rational agent, in virtue of engaging in action, is logically committed to accept a supreme moral principle, the *Principle of Generic Consistency* that commits every agent to respect the rights of freedom and wellbeing of all other agents including their own. The basis of his thesis is found in his doctrine that action has an inherent normative structure whose necessary features are freedom and wellbeing, and because of this structure every rational agent, just in virtue of being an agent, is committed to certain necessary prudential and moral constraints and in particular respect for all agents' rights to freedom and well-being.

11. A full and detailed defense of the argument for the PGC against all the major objections raised against it by various philosophers can be found in Spence 2006 (Chapters 1–3), Deryck Beyleveld (1991), and Alan Gewirth (1978).

12. Surprisingly, very little has been written on the relationship between information, knowledge, and wisdom in the philosophical literature, specifically with regard to information ethics. The main sources I will refer to in this part of the paper, but not exclusively, are Maxwell, N (2007); Tiberius, V. 2008; Vitek, B. and Jackson, W. (2008), Varelius, J. (2004); Kvanvig, L.J. (2003); Finnis (1983 and 1980); Ryan, S. (2007; 1999); Kekes, J. (1995); and generally the writings of Plato (*Apology*), Aristotle (*Nichomachean Ethics*) and the Hellenistic Philosophers (Epicureans, Sceptics, and Stoics) – for a discussion of those see (Spence, 2006, Ch. 10). See also, Vallor, S. (2016). *Technology and the Virtues: A Philosophical Guide to a Future Worth*. Wanting, Oxford: Oxford University Press.

13. For the purposes of this chapter, I will use the notions of *eudaimonia* and *eudaimonic* pluralistically as being potentially compatible with various different theories of the good life, including hedonistic, desire-satisfaction and objective-list theories, among others. Simply put, the notion of a *good life* used in this chapter is a good life that is in principle capable of leading to the attainment of eudaimonia. As such, any theory of a good life capable of leading to eudaimonia can at least in theory and upon further demonstration be considered a *eudaimonic* life. My own theoretical preference is a neo-Stoic eudaimonic life (as illustrated in chapter 3 of this book) that includes the virtues but that need not exclude other theories capable of also leading to the attainment of eudaimonia. See, for example, Vallor (2016). *Technology and the Virtues*.

14. See chapter 3 of this book that demonstrates how Gewirth's notion of self-fulfillment, as capacity fulfillment, is conceptually similar to the Stoic notion of eudaimonia.

15. Brink attributes the Eudaimonist axiom to Gregory Vlastos in Vlastos (1991) *Socrates: Ironist and Moral Philosopher*. Ithaca: Cornell University Press, p. 203.

16. For a detailed discussion of the Socratic elenchus and Socratic wisdom, see Hugh H. Benson (2000). *Socratic Wisdom: The Model of Knowledge in Plato's Early Dialogues*. Oxford University Press.

17. Stephanie Rice. https://www.smh.com.au/sport/swimming/i-want-you-to-know-how-sorry-i-am-tearful-rice-20100908-150s3.html. Accessed: April 25, 2021

18. Joel Monaghan. https://www.smh.com.au/sport/nrl/joel-monaghan-in-tears-after-quitting-the-raiders-20101109-17lff.html. Accessed: April 25, 2021

19. Catherine Deveny. https://www.smh.com.au/entertainment/tv-and-radio/deveny-dropped-as-columnist-for-the-age-20100504-u6si.html. Accessed: April 25, 2021.

20. Nir Rosen. https://www.news.com.au/world/journalist-nir-rosen-quits-after-insensitive-and-offensive-lara-logan-tweets/news-story/356d46509da0ee10a5d705c226087ee6. Accessed: April 25, 2021.

21. Elliott and Spence, *Ethics for a Digital Era*, Chapter 10; and Spence, *Wisdom and Well-being in a Technological Age*.

22. For the term *prudence see Note 6 above*.

23. See for example, Christopher Burr, Mariarosaria Taddeo and Luciano Floridi (2020). "The Ethics of Digital Well-Being: A Thematic Review," *Science and Engineering Ethics*, 26:2313–2343; Vallor. (2016). *Technology and the Virtues*.

REFERENCES

Annas, Julia. 2011. *Intelligent Virtue*. Oxford: Oxford University Press.
Aurelius, Marcus.1964. *The Meditations,* translated by Maxwell Staniforth. Harmondsworth: Penguin Books.
Aristotle.1941. Nichomachean Ethics, in Richard McKeon (ed.), *The Basic Works of Aristotle.* New York: Random House, 935–1112.
Banicki, K. 2009. The Berlin Wisdom Paradigm: A Conceptual Analysis of a Psychological Approach to Wisdom, *History and Philosophy of Psychology*, 11(2): 25–36.
Beyleveld, Deryck. 1991. *The Dialectical Necessity of Morality: An Analysis and Defense of Alan Gewirth's Argument to the Principle of Generic Consistency.* Chicago: University of Chicago Press.
Burr, Christopher, Taddeo, Mariarosaria and Floridi, Luciano. 2020. "The Ethics of Digital Well-Being: A Thematic Review," *Science and Engineering Ethics*, 26:2313–2343.
Descartes, R. 1979. *Principles of Philosophy*, in *Philosophical Works of Descartes*, E. Haldane and G. Ross (trans. and eds.), London: Cambridge University Press, pp. 201–302.
Dretske, Fred. 1999. *Knowledge and the Flow of Information.* Stanford: CSLI Publications.
Elliott, Deni, and Spence, H. Edward. 2018. *Ethics for a Digital Era*. Oxford: Wiley-Blackwell, Chapter 10.
Erasmus, D. 1994. *In Praise of Folly*, A. H. T. Levi (ed.) and B. Radice (translator). UK: Penguin Classics.
Finnis, J. 1983. *Fundamentals of Ethics.* Oxford: Clarendon Press Oxford.
———. 1980. *Natural Law and Natural Rights.* Oxford: Clarendon Press Oxford.
Floridi, Luciano. 2005. Is Semantic Information Meaningful Data? *Philosophy and Phenomenological Research*, LXX:2, 351–370.
———. 2002. What is the Philosophy of Information? *Metaphilosophy*, 33, 123–145.
———. 1999. Information Ethics: On the Theoretical Foundations of Computer Ethics. *Ethics and Information Technology*, 1(1), 37–56.
Garrett, R. 1996. Three Definitions of Wisdom. In Lehrer et al. (eds.), *Knowledge, Teaching, and Wisdom.* Dordrecht: Kluwer Academic Publishers, pp. 221–232.
Gewirth, Alan. 1996. *The Community of Rights.* Chicago, University of Chicago Press.
———. 1998. *Self-fulfillment.* Princeton, NJ: Princeton University Press.
———. 1978. *Reason and Morality.* Chicago: University of Chicago Press.
Kekes, John. 1995. *Moral Wisdom and Good Lives.* Ithaca: Cornell University Press.
Kekes, John. 1983. Wisdom, *American Philosophical Quarterly*, 20(3): 277–286.
Kvanvig, J. L. 2003. *The Value of Knowledge and the Pursuit of Understanding.* Cambridge: Cambridge University Press.
Kunzmann, Ute and Baltes, Paul. 2005. The Psychology of Wisdom: Theoretical and Empirical Challenges, in Robert J. Stenberg and Jennifer Jordan (eds.), *A Handbook of Wisdom: Psychological Perspectives.* New York: Cambridge University Press.

Maxwell, N. 2007. *From Knowledge to Wisdom* (Second Edition), London: Pentire Press.

Nozick, R. (1989). What is Wisdom and Why Do Philosophers Love It So? *The Examined Life*. New York: Touchstone Press, pp. 267–278.

Plato. 1978. The Apology, in *The Collected Dialogues of Plato*, Edith Hamilton and Huntington Cairns (eds.). Princeton: Princeton University Press, pp. 575–844.

Ryan, S. 2007. Wisdom. *Stanford Encyclopedia of Philosophy*. Accessed June 14, 2010.

Ryan, S. 1999. What is Wisdom? *Philosophical Studies,* 93: 119–139.

Spence, Edward H. 2020. The sixth estate; tech media corruption in the age of information. *Journal of Information, Communication and Ethics in Society,* 18(4):553–573. Bingley, UK: Emerald Publishing Limited, DOI: 10.1108/JICES-02-2020-0014.

Spence, Edward H. 2009. A Universal Model for the Normative Evaluation of Internet Information, *Ethics and Information Technology*, 11(4): 243–253.

Spence, Edward H. 2011. Information, Knowledge and Wisdom: Groundwork for the Normative Evaluation of Digital Information and Its Relation to the Good Life, *Ethics and Information Technology*, 13(3): 261–275.

Spence, Edward H. 2006. *Ethics Within Reason: A Neo-Gewirthian Approach*. Lanham: Lexington Books (a division of Rowman and Littlefield).

Kunzmann, Ute and Baltes, Paul B. 2005. The Psychology of Wisdom: Theoretical and Empirical Challenges. In Sternberg, Robert J. and J Jordan, Jennifer (eds.), *A Handbook of Wisdom: Psychological Perspectives*. Cambridge: Cambridge University Press.

Tiberius, V. 2008. The *Reflective Life: Living Wisely With Our Limits*. Oxford: Oxford University Press.

Vallor, S. 2016. *Technology and the Virtues: A Philosophical Guide to a Future Worth* Wanting, Oxford: Oxford University Press.

Varelius, J. 2004. Objective Explanations of Individual Well-Being. *Journal of Happiness Studies* 5: 73–91.

Vitek, B. and Jackson, W. 2008. *The Virtues of Ignorance: Complexity, Sustainability, and the Limits of Knowledge*. Lexington, Kentucky: The University Press of Kentucky.

Chapter 6

Tech Media Corruption in the Age of Information

The fault dear Brutus is not in our stars but in ourselves

—Shakespeare, Julius Caesar,
Act I, Scene III, L. 140–141.

INTRODUCTION

The primary aim of this chapter is to examine if digital information created, disseminated, mediated and curated increasingly via the Big Tech companies, such as Facebook and Google, are also subject to those same normative principles as the 4th and 5th Estates.[1] Having established that Facebook, as an illustrative example, is in essence a *media company*, the second part of the chapter will demonstrate, on the basis of the *Dual Obligation Information Theory (DOIT)* examined in chapter 5, that Facebook is also subject to the same normative principles and requirements that other media companies of the 4th and 5th Estates, are. The third part of the chapter will examine if Facebook's role in the *Cambridge Analytica* case not only violated the fundamental normative principles and requirements to which all media companies are subject, but moreover, its actions constituted media corruption.[2] More generally, Facebook's media corruption as illustrated in the Cambridge Analytica case, relates to a conflict of interest emanating from its Business Model that in its design and practice is conducive to systemic media corruption (Spence, 2020).

Recent publications by scholars such as Shoshana Zuboff (2019) and Frank Pasquale (2015) among others, as well as several articles by investigative journalists of legacy newspapers such as the *New York Times*, the *UK*

Times, *The Guardian*, the *Washington Post*, the *Wall Street Journal* and the *Sydney Morning Herald*, as well as the *Conversation*, among others, provide persuasive evidence that such practices are at least illegal and unethical. Do such practices, however, also constitute corruption and specifically, media corruption?

Digital media and by extension social media, mediated via Facebook and other tech companies such as Google challenge the traditional role and legitimacy of the traditional legacy media in the form of the 4th Estate, and by extension the digital media in the form of the 5th Estate, which includes user-generated information via the social media (Elliott and Spence, 2018, Chapter 6). Until the advent of the Big Tech companies, such as Facebook and Google, the 4th Estate and the 5th Estate constituted, as the *Convergent Media*, the primary sources of information communicated to the public (Elliott and Spence 2018, Chapter 4). Digital media now available through the Internet, Twitter, smart phones and social networks and platforms, such as Facebook, Instagram, Google and YouTube, are having a profound impact on the social life and democratic processes and practices of societies.

The 5th Estate is defined generally as the media estate comprising all the world denizens operating in cyberspace that as individuals or groups, disseminate information on matters of public interest to the world at large and some who are doing so with journalistic intent in their capacity as citizens – journalists of the 5th Estate. The *Convergent Media* referred to in this chapter comprises the symbiotic relationship that has developed between the journalists of the legacy media (the 4th estate) and the digital media including the social media (the 5th estate). Many journalists now operate in both those estates. When the symbiotic relationship between the two media estates operates appropriately and in accordance with the normative requirements of the DOIT that we examined in chapter 5, it augments the quality and quantity of information disseminated to the public and enhances the substance and scope of deliberative democracy (Elliott and Spence 2018, Chapter 6).

Truth, trust and the public's right to information are fundamental principles of the media professions and particularly journalism in their role as the 4th Estate and by extension the 5th Estate.[3] Those principles are mandatory and required by DOIT which was specifically designed (Spence 2011, 2009) to identify, analyse and evaluate normative issues that arise in the media generally on the basis of what information is intrinsically, and the normative implications of its different modes of communication in both the 4th and 5th Estates. This chapter comprises three main primary objectives:

Choosing *Facebook* as the specific example of what is referred to in this chapter as the *6th Estate* to distinguish it conceptually from the 4th and 5th Estates, the primary objective of the first part of the chapter is to examine if digital information created, disseminated and mediated increasingly via

Tech Companies such as Facebook is also subject to those same normative principles as the 4th and 5th Estates. If it is, should we not be consistent and extend the role of the media, as creators, curators, mediators and communicators of information to include Tech Media companies, such as Facebook?[4]

Having established in the first part of the chapter that Facebook is in essence a *media company*, the second part of the chapter will demonstrate, on the basis of the DOIT, that Facebook and generally by analogy other tech companies, such as Google, are also subject to the same normative principles and requirements that other media companies of the 4th and 5th Estates, are. This will be illustrated through an examination of Facebook's media role in the Cambridge Analytica case.

The third part of the chapter will examine if Facebook's role in the Cambridge Analytica case not only violated the fundamental normative principles and requirements to which all media companies are subject, but moreover, its actions constituted media corruption. More generally, Facebook's media corruption as illustrated in the Cambridge Analytica case, relates to a conflict of interest emanating from its Business Model that in its design and practice is conducive to systemic media corruption.

Finally, the chapter will analyse the *corrupting effects* of Facebook's actions in the Cambridge Analytica case on digital information and its communication, which undermine democracy and its underlying political institutions as well as the rights of citizens.

WHAT IS FACEBOOK'S PROFESSIONAL AND INSTITUTIONAL ROLE?

Before we proceed to enquire if and how Facebook is also committed to the same fundamental normative principles as other media companies, such as truthfulness and trustworthiness and the public good, we must first establish what the professional role of Facebook, in its capacity as an information and communication technology company, is. Is the professional role of Facebook that of a mediator of its user's information? Custodian of its users' information? Curator of its users' information? Trustee of its users' information? Trader of its users' information? Is Facebook essentially, all things considered, a media information/communication company, or something else entirely, and if so, what is its overarching professional and institutional role? After all, a lot of information and communication of information is channelled through Facebook by its billions of information users daily. Prima facie at least, it seems that Facebook is by its function and design if not by name, operating as a media company. If so, Facebook is normatively responsible for the information and communication it handles and should be held both epistemically, ethically and eudaimonically responsible and accountable as other media companies are. Why is it not?

Facebook is shy about letting the rest of us know exactly how they themselves perceive its respective professional and institutional role, content to inform us that they are merely a "platform." Which is convenient as the term "platform" seems topic-neutral and free of any normative commitments with regard to the dissemination and communication of information via Facebook. Which, would appear to suit Facebook as it alleviates the company of any normative commitments as those that apply to the convergent media of the 4th and 5th Estates and generally to anyone that disseminates and communicates information to others on the World Wide Web, on the basis of the DOIT-Wisdom model.

As we examined earlier in chapter 5, DOIT locates the normativity of the communication of information within the inherent normative structure of information that comprises epistemic, ethical and eudaimonic features (Spence, 2011, 2009). Those normative features of information commit all communicators of information generally, and especially media companies, including Facebook qua media company, to principles of truthfulness, accuracy, reliability, trustworthiness and the public good. Furthermore, as illustrated in section, Does Facebook's Part in the Cambridge Analytica Case Constitute Media Corruption?, through discussion of the Cambridge Analytica case, the violation of those fundamental media principles by Facebook may also constitute media corruption. As we shall see, that systematically occurs when its Business Model of generating advertising revenue on the basis of users' content, comes into direct conflict with its media role, which results in the violation of its normative commitments and responsibilities, as required by DOIT.

On closer examination, and regardless of its self-describing professional role as a "platform" Facebook appears to be a media company that handles, curates, mediates and communicates information generated by its users for profit via advertising and marketing and as such should be viewed in essence as a media company. As a media company, Facebook is then normatively responsible and accountable to its users and the public generally, as other media companies are, including radio, television, cinema, video, newspapers, magazines, books and all other types of media broadcasts and podcasts, as well as publications that generate, disseminate, exchange, mediate and communicate by analogue and digital channels information in the public domain. Even if we accept the claim that Facebook is merely operating as a "platform," it is still normatively responsible and accountable for what occurs on its information communication platforms. According to Scott Galloway (Galloway 2017,117),

> Facebook attempts to skirt criticism of its content by claiming it's *not* a media outlet, but a *platform*. This sounds reasonable until you consider that the term

platform was never meant to absolve companies from taking responsibility for the damage they do. What if McDonald's, after discovering that 80 percent of their beef was fake and making us sick, proclaimed they couldn't be held responsible, as they are not a fast-food restaurant but a fast-food platform? Would we tolerate that?

The point Galloway is making here is that the mere description of what Facebook does as a "platform" does not absolve that company of responsibility and accountability for any illegal or unethical transgressions that occur on its platform.

To emphasise again the direct relevance of DOIT's normative implications for Facebook's operations we shall examine in more detail in section, The Cambridge Analytica Case, DOIT locates the normativity of the communication of information within the inherent normative structure of information that comprises epistemic, ethical and eudaimonic features. These normative features of information commit all communicators of information generally, and especially media companies, including Facebook qua media company, to mandatory principles of truthfulness, accuracy, reliability, trustworthiness and the public good. Furthermore, as illustrated in section, Does Facebook's Part in the Cambridge Analytica Case Constitute Corruption?, through discussion of the Cambridge Analytica case, the violation of those fundamental media principles by Facebook may also constitute media corruption. As we shall see, that occurs systematically, when Facebook's *Business Model* comes into direct conflict with its *Media Role*, which results in the violation of its normative commitments and responsibilities required on the basis DOIT. We shall see more clearly how DOIT dovetails the media corruption model presented in this chapter and one we examined in detail in chapter 5; both DOIT and the media corruption model working in tandem reveal the media corruption that is at the heart of Facebook's business operations.

Let us now explore in more detail what is the actual role of Facebook, as calling it a platform does (a) not clearly define or explain what its overarching role is and (b) calling it a "platform" does not absolve the company of any legal or ethical responsibility in regard to its practices. Facebook is the medium through which approximately 2.2 billion people worldwide (Galloway 2017, 90) disseminate and communicate text and visual information, using *Facebook*, *WhatsApp* and *Instagram*. Facebook likes to think of and define itself as a *platform*, but as we saw calling itself a platform does not absolve that company of any normative responsibility but merely obfuscates what Facebook's actual role is, which if examined carefully in terms of what Facebook actually does, is that of a *media company*. For Facebook collects, stores, curates, mediates as well as moderates the information of its users, not unlike what newspapers such as the *New York Times*, *the UK Times* and *The*

Guardian do with information provided by their subscribers through comments posted on their respective newspaper "platforms." The extra activity, in which Facebook engages, is to sell its users' information for profit to advertisers and other information contractors, often without the knowledge and consent of its users, as in the case of *Cambridge Analytica*. However, that is also not entirely new, as some traditional media companies in the form of newspapers, commercial radio and TV stations have done something similar by presenting paid for ads and PR media releases disguised as editorial comment to its audiences, generally known in the industry as advertorials. Such advertorials not only violate epistemic and ethical principles but area also conducive to media corruption (Spence 2021; Spence 2008; and Spence et al. 2011).

Similar in kind but expressed differently in practice and to a much greater extent, that is what Facebook's business model essentially is: collecting and selling their users' information to advertisers and other information brokers for profit, often without the explicit knowledge and consent of their users, as in the case of *Cambridge Analytica*. The main difference between what Facebook does and how traditional media use advertorials is that the content used by Facebook is provided by its users and surreptitiously sold to advertisers. Both, however, the media audiences and the Facebook users are used as advertising target practice.

On the above analysis so far, Facebook appears for all intents and purposes to operate as a *media company* for which its 2.2 billion users provide the content. A very valuable content, which Facebook sells to advertisers and other information brokers to make Facebook, along with Apple, Amazon, Microsoft and Google, one of the wealthiest companies in the World. In 2018, Facebook had a US$534bn market capitalisation (Galloway 2017, 7). Nice work if you can get it and get away with it!

So, having established that Facebook in virtue of its primary role of collecting, storing, curating and mediating information content generated by its users is a media company, if not by name by activity at least, it should be held to the same normative standards and responsibilities as other media companies are. Let's not forget that as we saw earlier in this chapter, the convergent media comprising the 4th and 5th Estates is also normatively held to account for user-generated content in its publications. By parity of argument, Facebook as part of what is referred to in this chapter as the *6th Estate*[5] should also be held to the same normative media standards. That requires Facebook to act responsibly in monitoring and editing information disseminated on its platform by its users as other media companies and publishers do, *Wikipedia*, for example. Though the information comes from its users, that does not alleviate Facebook of its primary responsibility of checking the normative quality and integrity of that information. Especially, regarding *fake news, misinformation and disinformation*.

Let us now examine more specifically what those normative standards, principles and responsibilities are vis-a-vis Facebook's role morality based on the DOIT and by extension the DOIT-Wisdom theory that was examined in chapter 5. Henceforth, unless otherwise specified, the term DOIT will be used for economy and convenience in this chapter.

FACEBOOK'S NORMATIVE COMMITMENTS AS A MEDIA COMPANY

As a media company, Facebook is committed, like all other media companies, to the fundamental principles of truth, trust and the public's right to be informed truthfully, reliably and transparently, based on those principles. This is simply because if Facebook is a media company that collects, curates and mediates communication of information then it is bound to the normative commitments that emanate from within the inherent normative structure of information as demonstrated by DOIT, which was examined in chapter 5.

Having established, in accordance with the normative requirements of DOIT, that Facebook as a media company, bound by the same normative principles and responsibilities as other media companies, Facebook's normative violations against the requirements of DOIT generally, express themselves in at least two primary ways in practice:

First, it violates the normative requirements of DOIT as a media company by its failure of not adequately monitoring and editing through prevention and correction, misinformation, disinformation and fake news on its platform. Facebook has denied that it has any normative responsibility for monitoring, fact checking and editing political advertising content placed by rival political parties on its platform (Milkman, 2019), thus, conveniently shirking any of its normative responsibilities as a media company, as required by DOIT. However, following Mark Zuckerberg's pronouncements on Facebook in 2019 (Zuckerberg, 2019) and in an op-ed article in the *Financial Times* in 2020 that "Big Tech Needs More Regulation" (Zuckerberg, 2020), that might change.

The second way in which Facebook violates the normative requirements of DOIT as a media company is by its harvesting of users' information with the aid of opaque algorithms and its business model, which it sells to advertisers and other information brokers for profit. And as we shall see in the case of Cambridge Analytica, it does so without the users' informed consent. This general practice, which as we shall examine in more detail in section, The Corrupting Effects of Facebook's Violations of it's Normative Media Responsibilities, involves a *conflict of interest* between its role as a media company and its business model whose overarching goal is to sell user's information for advertising profit, without its users' informed consent, as illustrated in the Cambridge Analytica case. This conflict of interest lies at the heart of Facebook's operations and as we shall examine in more detail below is conducive to systemic media corruption. Suffice to say in this section, that this second expression of Facebook's systemic violation of its normative responsibilities as a media company, as required by DOIT, also reflects a form of media corruption also present in the traditional media legacy companies representative of the 4th Estate, as well as the 5th Estate. The type of media corruption is in the form of *advertorials*: the practice of presenting paid-for-content in the form of advertisements or press-release material deceptively, as editorial or news content (Spence, 2008).

In the case of traditional advertorials by the media companies of the 4th and 5th Estates, the content is generated by those companies in partnership with the advertisers or PR companies and presented to their audiences as credible and objective journalistic content or editorial comment, without the users being aware of the deception and manipulation. That practice is sometimes referred to as an "unholy alliance" or "media corruption by collusion" between journalism, advertising and public relations (Spence 2018, 2017, 2014, 2008; Spence et al., 2011). Interestingly, in the case of Facebook, this process is reversed. The "advertorials" are created through users' content without their knowledge or informed consent, and then sold to advertisers, who, in turn, target the users with advertisements, tailored specifically to the users' individual profiles created by Facebook's algorithms. As we shall see in section, The Normative Violations by Facebook in the Cambridge Analytica Case, in the case of Cambridge Analytica, a similar advertising technique was used in 2016 to influence the U.S. presidential elections as well as the UK Brexit referendum.

Although the process involving traditional advertorials in the media of the 4th and 5th Estates, and Facebook's surveillance advertising practice based on its business model are different in type, they are similar in kind. Both are designed to deceive, manipulate and exploit their users through misinformation. Another characteristic they both have in common is a *conflict of interest*. A conflict between a media company's professional role morality of disseminating and communicating information to the public, and in Facebook's case, its role morality as a media company of mediating and curating its users' information, held for them in trust, and on the flip side of the conflict, Facebook's financial interest in the form of advertising profits, specifically in situations when the latter interest is allowed to come into conflict and take priority, over the former interest. As we shall examine in more detail in section, The Corrupting Effects of Facebook's Violations of its Normative Media Responsibilities, such conflicts of interest involving a conflict of roles, expressed either in the form of advertorials, or in the form of surveillance targeted advertising via profiling, as in the case of Facebook and analogously

Google, are conducive to media corruption as well as violate the epistemic and ethical principles required by DOIT. As we can see, there is clearly a close conceptual relationship between the normative requirements of DOIT and those of the Media Corruption model that is explained by a common normative denominator they share, that of morality. This helps explain why media corruption and corruption generally, is *fundamentally a moral problem* (Miller, 2018, Miller, Roberts, and Spence, 2005).

Having examined Facebook's two general types of normative violations with regard to the normative principles required by DOIT, and initially at least with regard to the media corruption model; and having established the close conceptual normative link between the two, we shall now examine more specifically Facebook's normative violations and failures in the Cambridge Analytica case. The purpose is to determine through an analysis of that case study, whether any of those normative failures also constituted media corruption and more generally, systemic media corruption.

THE CAMBRIDGE ANALYTICA CASE

One of the biggest and most controversial cases involving Facebook was the Cambridge Analytica case[6] that exposed the sale of millions of users' data to third-party information brokers without the users' knowledge or consent. As a result of that exposure, Facebook was fined $US5 billion as part of a settlement with the Federal Trade Commission (USA Today: 2019) and £500, 000 GBP by the UK Information Commissioner's Office (ICO). What is of interest about this case and other similar cases involving both Facebook and Google as well as other tech companies is the role the 4th and 5th Estates played in exposing the unethical and illegal practices of those companies. It's a case of the convergent media of the 4th and 5th Estates keeping a check on the 6th Estate that comprises these new tech media companies.

In summary, as described in Wikipedia (see note 6), the Cambridge Analytical case is as follows:

> In March 2018, multiple media outlets broke news of Cambridge Analytica's business practices. *The New York Times* and the *Observer* reported that the company had acquired and used personal data about Facebook users from an external researcher who had told Facebook he was collecting it for academic purposes. Shortly afterwards, the BBC Channel 4 aired undercover investigative videos showing Nix [Cambridge Analytical CEO Alexander Nix] boasting about using prostitutes, bribery sting operations and honey traps to discredit politicians on whom it conducted opposition research and saying that the company "ran all of (Donald Trump's) digital campaign." In response to the media reports, the Information Commission of the UK pursued a warrant to search the

company's servers. Facebook banned Cambridge Analytica from advertising on its platform, saying that it had been deceived. On 23 March 2018, the British High Court granted the Information Commissioner's Office a warrant to search Cambridge Analytica's London offices. The personal data of up to 87 million Facebook users were acquired via the 270,000 Facebook users who used a Facebook app called "This Is Your Digital Life. By giving this third-party app permission to acquire their data, back in 2015, this also gave the app access to information on the user's friends network; this resulted in the data of about 87 million users, the majority of whom had not explicitly given Cambridge Analytica permission to access their data, being collected. The app developer breached Facebook's terms of service by giving the data to Cambridge Analytica. On 1 May 2018, Cambridge Analytica and its parent company filed for insolvency proceedings and closed operations. (https://en.wikipedia.org/wiki/Cambridge_Analytica)

A key question of this case and other similar cases involving the collection and distribution of users' data to unknown third parties to be used for unknown purposes, was whether Facebook had obtained *informed consent* from its 87 million users whose data they collected and passed on to Cambridge Analytical. It transpires that only 270, 000 Facebook users gave their consent by installing an app that allowed Facebook to collect information from their Facebook profiles and those of their friends' profiles, which they then provided to the makers of the app who, in turn, passed it on to Cambridge Analytical. But apart from the 270, 000 Facebook users who had installed the app and thus could be presumed to have given their consent, though not necessarily for the purpose for which it was used, most of the remaining 87 million Facebook users whose information was harvested for use by Cambridge Analytical, were unaware their data had been collected in this way and that it would be used to form voter targeting for Trump's presidential election campaign (Tufekci Zeynep, 2018). So, most of those Facebook users not only did not give consent for the harvesting of their information by Facebook, but moreover did not provide consent for the purpose of its use, that is, to target voters in the Trump presidential election. Because of those reasons Facebook was found guilty of breach of user's privacy and fined by the Federal Trade Commission, as well as by the UK ICO.

THE NORMATIVE VIOLATIONS BY FACEBOOK IN THE CAMBRIDGE ANALYTICA CASE

Having determined the key facts of the Cambridge Analytical case let us now examine the specific normative issues of this case by application of DOIT.

Facebook's part in allowing the harvesting of its user's information for use by Cambridge Analytica was an epistemic and ethical violation of their *right to informed consent*, their *right to privacy* and crucially, by implication, their *right to autonomy*, as they were offered no knowledge that their information was collected and sent to a third party, and furthermore this was done without their knowledge of the purpose for which their information was to be used by Cambridge Analytica. In doing so, Facebook violated the fundamental communication principles of trust, reliability and truthfulness, as well as its users' rights to freedom and well-being, which supports their fundamental right to autonomy as informational communicative agents (Gewirth 1978, Spence 2006).

Moreover, as we examined previously in The Cambridge Analytica Case Facebook's actions in the form of an "advertorial" of selling users' information for advertising profit without their knowledge and consent, similarly as in the case of traditional advertorials used by the legacy media, Facebook's use of "advertorials" not only constituted normative violations of the requirements of DOIT, but in addition, it also constituted media corruption, a topic we shall examine in more detail in sections (8 and 9). To distinguish the type of advertorials used in legacy media from the type used by Facebook, I shall refer to Facebook's use of their type of "advertorials" as *advert-facetorials*, a neologism I specifically use in this chapter to distinguish Facebook's surveillance type of "advertorial" from traditional advertorials by legacy media.

In sum, Facebook's actions in this case at least, both by omission and by commission, were unethical. To further investigate how Facebook's actions in the case of Cambridge Analytica constituted media corruption, and whether this was a one-off instance of corruption or a case of systemic corruption, we will now examine what the characterising features of corruption as well as the necessary and sufficient conditions of corruption generally and media corruption specifically are. This will be done by means of an examination of the Myth of Gyges in Plato's *Republic* (Spence, 2008).

THE MYTH OF GYGES AND THE CHARACTERISING FEATURES OF CORRUPTION

Once upon a time, a certain shepherd from Lydia named Gyges, while tending to his sheep, found a ring. He soon discovered that by turning the ring on his finger, he could make himself invisible. A few days later, he went to the palace with a delegation of shepherds to see the king. By making himself invisible, he seduced the queen, killed the king and assumed total power by becoming the king himself. Another character, Glaucon, continues the story by asking us to imagine an ordinary person who, like Gyges, has the ability through possession of a similar magical ring to render him or herself invisible. Invisibility would allow that person the opportunity to act unethically at will with total impunity.

Glaucon refers to this possibility as the "highest reach of injustice, being deemed just when you are not." He labels this "the most perfect injustice" (Plato 1952). Given such a possibility, what possible reason would someone like Gyges have for not being corrupt but being non-corrupt and moral instead? The question of "why be moral" under Gygean conditions of perfect injustice is referred to in the ethics literature as the "authoritative question of morality" (Spence 2006; Spence et al. 2011; Gewirth 1978).

Five features emerge that seem, at least initially, to characterise corruption: the *possession of power*, a *disposition* to exercise that power, an *opportunity* to exercise that power, *invisibility* or *concealment*, and *self-regarding gain*. We will define *power* as the possession of the ability or capacity to act in a manner capable of bringing about a certain intended desired outcome; a *disposition to exercise that power*, as the possession of a pre-disposition, pro-attitude or willingness to purposefully exercise that power; o*pportunity*, as having the opportunity, either presented to oneself or engineered by oneself, to engage in some activity for which one has the power and the pre-disposition in which to engage; *invisibility* or *concealment*, as the ability or quality an agent has for keeping the motives and the identity of the agency of his actions invisible, concealed or hidden from the gaze of others. *Self-regarding gain* is any gain, not necessarily financial, which accrues to the agent personally or to a group of which he is a member, as a result of his or the group's actions. With regard to the condition of invisibility, while the actions themselves would be visible at least with regard to their effects and consequences, the identity of their agency is invisible. Or at least the corrupt agent's intention is to keep his agency invisible or concealed.

The five characterising features of corruption are almost always present and accompany acts of corruption. For without power, understood as the ability to act in a manner capable of bringing about a certain intended desired outcome, one cannot commit a corrupt act. Without a disposition to exercise that power willingly, the possession of power to engage in corrupt activities may not be exercised. Similarly, through lack of opportunity one cannot engage in corrupt activity even when one has both the power and the disposition to act corruptly. Invisibility seems to also be a characterising feature of corruption that is usually, if not always, present in instances of corrupt activity.

Invisibility seems to be at least instrumentally desirable,[7] for without invisibility one might not be able to evade detection, thus escaping possible social disapproval and punitive retribution from others or the state. Even for a Gyges-like person, keeping the identity of his agency in committing the immoral acts hidden, lest he invoke social disapproval which may eventually undermine his power to rule and invite retribution from those he harms through his unethical conduct, would be prudent. Like Glaucon's perfectly unjust person, perfectly corrupt persons maintain an outward appearance of probity, justice and morality, while carrying out their corrupt deeds in secret. In this way, they maximise their self-regarding gain, which can either accrue to them personally or to a group to which they belong or cause to which they are committed, with little or no instrumental cost to themselves, their group or their cause.[8]

Sometimes the corruption of one's deeds and the corruption of one's character may remain invisible even to the corrupt agent themselves. The invisibility may be self-induced through self-justifying rationalisations, and ignorance may be manifested and expressed as lack of self-reflection and self-knowledge. Usually, self-justifying rationalisations and ignorance are closely related. For often it is ignorance – a lack of self-examination and self-reflection – that necessitates self-justifying rationalisations on the part of the corrupt agent. It is plausible to argue that corruption and unethical conduct may at times remain invisible to the corrupt agents themselves through a dispositional or actual lack of self-examination and self-reflection.

If all the above five features are regular features that normally accompany typical cases of corruption, they are nevertheless not sufficient in fully characterising corruption. They are not sufficient, for if they were, a house burglar or a professional bank robber would be deemed corrupt. However, though immoral, the actions of the house burglar and bank robber are not what we would normally describe as corrupt. The missing sufficient condition is a socially pre-established and widely acknowledged fiduciary relationship of trust that exists between the corrupt person or corrupt group and the person or group of persons who are harmed in some way by the corrupt person's or the corrupt group's actions.

The reason why house burglars or bank robbers are not normally deemed corrupt is because there is an absence of a prior fiduciary relationship of trust between the burglar and the bank robber on the one hand, and those who are harmed by their actions on the other; namely, the household owners, the banks and their customers. The addition of the condition of a fiduciary duty is in keeping with one of the traditional dictionary definitions of "corruption," namely, "the changing from the naturally sound condition" or "the turning from a sound into an unsound impure condition" or "the perversion of anything from an original state of purity."[9] So, violation of fiduciary duty is the sixth characterising feature of corruption.

The fiduciary relationship can be articulated in political, professional, institutional, social or familial terms. Thus, priests, ministers of religion or teachers who sexually abuse children in their care, are acting corruptly, even though there might not be any financial gain, as in the case of financial fraud.

DOES FACEBOOK'S PART IN THE CAMBRIDGE ANALYTICA CASE CONSTITUTE MEDIA CORRUPTION?

All six features and their enabling conditions that characterise corruption and media corruption specifically seem to be present in the way that Facebook acted in the Cambridge Analytica case. Namely, *possession of power*, a *disposition* to exercise that power, an *opportunity* to exercise that power, *invisibility* or *concealment*, and *self-regarding gain* and abuse of a *fiduciary relationship of trust*. Facebook had the power as well as the disposition and opportunity to exercise that power and with the exception of 270, 000 people, did so under concealment without informing 87 million other users and without obtaining their informed consent, for the purpose of collecting and passing on those users' information to Cambridge Analytica for a self-regarding gain and by doing so, violating the fiduciary duty of trust owed to and expected by their users. As such, we can conclude that Facebook's actions by commission and omission in the case of Cambridge Analytica, had all the six characterising features of media corruption.

This assessment also supports our earlier analysis in section, The Cambridge Analytica Case, where we discussed the practice of advertorials as a form of media corruption. In the Cambridge Analytica case, Facebook's role in allowing the harvesting of users' content to be used for targeted political advertising, as in the 2016 presidential elections and similarly in the Brexit Referendum in the UK, constituted media corruption in the form of *advertorial*. As mentioned earlier in section (4), this is a Facebook type of "advertorial" based on users' content served to them in the form of "surveillance targeted advertising," without the users' knowledge and informed consent. Notice how an advertorial, has all the characterising features of media corruption and corruption generally, and as we shall examine in section, The Corrupting Effects of Facebook's Violations of its Normative Media Responsibilities, it also involves a conflict of interest, which is in most cases, conducive to media corruption and corruption generally. This further illustrates that such conflicts of interest are particularly ethically problematic as they tend to give rise to systemic corruption, as in the case of Facebook. Another observation worth reiterating at this point is that Facebook's role in the Cambridge Analytica case also illustrates the close conceptual and normative relationship between DOIT and media corruption model used in this book. Namely, that Facebook's normative violations against the requirements of DOIT were pre-conditions for further violations in the form of media corruption. That is because media corruption, as in the case of Facebook's violations of DOIT as canvassed above, is also fundamentally an epistemic and ethical problem, a problem of trust.

Facebook may raise the objection that they did not know that the 270, 000 users who gave their consent would pass on the information to their friends and so on. Even if that were the case, doubtful as that might seem, given that Facebook has a good knowledge of how their system works and how it is used (the more clicks by more users the better financially it is for Facebook) then at the very least Facebook acted negligently. That still amounts to media corruption by negligence as it involves a failure of the normative responsibilities they owed to their users and a failure of the fiduciary duty of trust to their users. By analogy it would be similar if a patient were harmed by the negligence of his medical doctor who negligently administered the wrong treatment, when the doctor knew or should have known the correct treatment.

THE CONFLICT OF INTEREST AT THE HEART OF FACEBOOK'S BUSINESS MODEL

The core normative problem for Facebook's operations that encompasses ethical, epistemic and eudaimonic requirements and responsibilities, as per the DOIT, is a *conflict of interest* between the financial interests of Facebook given its business model on the one hand, and the normative interest of its users, given Facebook's role morality as a media company, on the other. Conflicts of interest can be conducive to corruption. Facebook's conflict of interest arises out of its business model that is based primarily on surveillance, the watching, tracking and collection of vast amounts of its users' data through sophisticated AI algorithms for the overarching purpose of selling their users' content to advertisers, political actors and other information brokers, such as in the Cambridge Analytica case. However, in the case of Cambridge Analytica, the normative interests of Facebook's users, epistemic, ethical and eudaimonic, are not only not served but undermined and violated. Facebook's users are merely the human resource in this unequal and seemingly unfair deal, which ironically provides the content that makes Facebook so successful and enormously wealthy.[10]

THE CORRUPTING EFFECTS OF FACEBOOK'S VIOLATION OF ITS NORMATIVE MEDIA RESPONSIBILITIES

What remains for us to examine now in this section more specifically, is what were the actual *corrupting effects* of Facebook's unethical and corrupt behaviour and who and what was corrupted and in what way, and what are the moral implications of that corruption? As the corruption involved is institutional corruption what we need to determine is what the corrupting effects were (a) on the institution itself by virtue of its role as a media company, (b) the institutional persons qua *institutional role occupants* or *participants*, in this case Facebook's users, (c) the other institutional actors involved, in this case the clients of Facebook, in their capacity as advertisers and other third parties, for example, Cambridge Analytica, to whom the users' information was sold and finally (d) the corrupting effect it had on democratic political institutions. Before we do, let us recall what constitutes corruption in terms of its *causal corrupting effects*.

As in the case of the characterising features of corruption referred to above, it would also be useful to examine in more detail the necessary and sufficient conditions of corruption, with regard to corruption generally and as they specifically apply to media corruption (Spence 2021; Miller 2018; Spence et al. 2015).

1. An instance of institutional corruption is an instance in which an action or joint action or set of actions performed by an agent or set of agents undermines or has a tendency to undermine or contribute to the undermining of a legitimate institutional role, process, goal(s) or purpose(s) (understood as a collective good) of that institution and/or an effect of the despoiling of the moral character of an institutional role occupant (s), qua role occupant of that institution. This is the *Institutional Actor Condition*.
2. An action or set of actions is corrupt only if it corrupts something or someone. This is the *Cause and Effect Condition*.
3. An action or set of actions is corrupt only if the role occupant(s) who performs that action(s) intended it or foresaw the corrupting effects of that action(s) – or at the very least, could and should have foreseen the corrupting effects – or could have reasonably avoided the action(s). This is the *Moral Responsibility Condition*.

A common trait of institutional corruption that is not, however, a *necessary condition*, is its predominantly *systemic* nature. It is not a necessary condition for even if the corruption as illustrated by the Cambridge Analytica was a one-off case, it would still count as corruption because of its high impact and wide unethical repercussions and damage to public trust.

Another general feature of corruption is that all forms of corruption are immoral, though not all forms of immorality constitute corruption. Many forms of corruption, such as political bribery, are illegal, though some, like fabricating stories in journalism and advertorials, are not (Spence et al. 2015). Corruption as understood and explained in this book and in this chapter specifically, is therefore fundamentally a matter of morality (Spence et al. 2015, Miller 2018). This is further illustrated by the close conceptual connection between corruption and unethical conduct as illustrated by the application of DOIT to the Cambridge Analytical case.

Three distinct types of institutional corruption we can identify with regard to Facebook's part in the Cambridge Analytica case, arising directly from its surveillance business model are, *information and communication corruption*, *democratic political corruption* and *citizenry corruption*, the latter being a type of *institutional personal corruption*.

Information and Communication Corruption

Applying the three conditions of the definition of corruption above, we can now demonstrate that Facebook's part in the Cambridge Analytica had a corrupting effect institutional media role in undermining the information and communication processes and purposes of the collection, mediation, curation and management of the information and its communication placed in trust on Facebook's site by its users, and as required by its fiduciary duty of trust as a media company (i.e., its *role morality*) and the normative requirements of DOIT that apply to the communication of information. This also demonstrates that corruption generally and media corruption specifically is fundamentally a moral problem, as demonstrated by the close conceptual relationship referred to above that holds between the DOIT and the media corruption model.

Another aspect of Facebook's media corruption as illustrated by the Cambridge Analytica case is its use of an *advert-facetorial*, a neologism I use in this chapter to distinguish Facebook's surveillance type of "advertorial" from traditional advertorials by legacy media. In the case of Facebook's use of advert-facetorials, users' information was harvested and used in political advertising in the 2016 U.S. presidential elections and also in the UK Brexit Referendum to influence voters. In addition, as previously canvassed, the use of advert-facetorials by Facebook, generated by its Business Model through non-transparent algorithms, which metaphorically, at least, parallel the "ring of invisibility" in Gyges Myth, further demonstrates that this form of media corruption that involves the selling of users' information to advertisers and other information contractors for profit without the users' informed consent, constitutes more generally, *systemic media corruption*.

Democratic Political Corruption

Applying the three conditions of the definition of corruption as above, we can demonstrate that Facebook's part in allowing Cambridge Analytical to target and use its 87 million users' profiles in Donald Trump's 2016 presidential campaign, had the effect of undermining the electoral process, which had the effect of also undermining the political institutions that comprise the democratic system itself. Even if Facebook had not intended their act of giving unauthorised access to their users' information for voter profiling as used by

Cambridge Analytica, they should have foreseen the corrupting effects this may have on the democratic process and taken steps to avoid the distribution of their users' profiles to Cambridge Analytica. As Zeynep Tufekci correctly observed in her *New York Times* article (March 19, 2018),

> A business model based on vast data surveillance and charging clients to opaquely target users based on this kind of extensive profiling will inevitably be misused. The real problem is that billions of dollars are being made at the expense of the health of our public sphere and our politics, and crucial decisions are being made unilaterally, and without recourse or accountability.

Tufecki's astute observation also seems to confirm our concern regarding Facebook's inherent and generic conflict of roles, to which, as discussed earlier, its surveillance business model gives rise and is thus conducive to systematic media corruption.

With regard to institutional political corruption as illustrated by the joint actions of Facebook and Cambridge Analytica, Dennis Thompson makes a point that is in keeping and lends support to our analysis on institutional corruption above. According to Thomson's understanding of institutional political corruption,

> The fact that an official [for our purposes, Facebook, and Cambridge Analytica corporate actors] acts under conditions that tend to create improper influence is sufficient to establish corruption, whatever the official's motive. (Thompson 2013, 13 in Miller 2018)

As in our analysis of conflicts of interest and conflicts of roles earlier, it is not the motive that is decisive in determining acts of corruption or at least acts and roles conducive to corruption but the *tendency to corrupt* in undermining a legitimate institutional purpose or role. According to Thompson, [a practice] is corrupt if it is of a type that tends to undermine [core] processes and thereby frustrate the primary purpose of the institution (Thompson 2013, 7 in Miller 2018).

A general conclusion we can derive from the above analysis so far is that the inherent conflict between Facebook's two roles that of the their surveillance business model and their role as a media company, has a tendency to undermine their media role by corrupting the purposes and processes of the information and communication of its users, as well as having a tendency to undermine the legitimate purposes and roles of the democratic political institutions, which those informational processes are meant to support. The result is a *perfect storm* and of great concern as the damage of corruption to our systems of information and communication, as well as to our democratic political institutions that are closely related through the power of social media, is substantial and ongoing.

Referring back to Plato's Myth of Gyges, we can now see more clearly that Facebook, as illustrated by the Cambridge Analytica case, acts like Gyges under a cloak of invisibility, acting corruptly for its self-interest and profit while simultaneously maintaining a superficial public profile of probity and legitimacy for the common good of humanity. What Plato refers to as the phenomenon of "perfect injustice, appearing just when you are not."

CONCLUSION: THE SYSTEMIC PROBLEM OF HARVESTING AND MARKETING INFORMATION

The problem of the collection and harvesting of users' information via opaque algorithms that is then sold for profit to advertisers, often without the users' knowledge and informed consent, is not limited to Facebook. Insofar as Google, like Facebook, operates as a media company in its information practices through its *search* function and its communication and curations practices that involve the collection, profiling and harvesting of users information via its apps such as Gmail, for example, which like Facebook sells to advertisers for profit; as well as its YouTube practices that have been partly shown to be responsible for the spread of misinformation, fake news and deep fakes, on the Internet generally and specifically during the U.S. presidential elections of 2016[11]; then Google's media role also places Google like Facebook, within the parameter of the 6th Estate. Therefore, some of the *types* of normative issues examined in this chapter with regard to Facebook, apply equally to Google; for example, its selection, collection and harvesting of users' information via its opaque algorithms, which then sells for profit to advertisers without their knowledge or informed consent. As in the case of Facebook, that problem is systemic: a conflict of interest between the collection and curation of users' information in its role of a media company, and its business model that is specifically designed and geared to generate massive profits through the sale of users' information to advertisers. And as in the case of Facebook, the more time users spend on Google search and more clicks on its other apps and more watch time on YouTube, the more money Alphabet-Google makes. That in itself is not a problem. The problem is the monetisation and weaponisation of the information of its users that gives Google enormous power and control to influence every aspect of our lives.

Operating like Gyges both within and outside the social contract for the common good,[12] Google operates under the concealment of its opaque AI-driven algorithms in near-monopolistic manner that allows it to evade and avoid the epistemic and ethical responsibilities of transparency, accountability, truthfulness and trustworthiness by which other media companies are bound. As in the case of Facebook's operations, illustrated by the Cambridge Analytica case, Google's operations undermine citizens' democratic rights to privacy, autonomy and well-being, as well as the democratic system and its associated institutions. The real problem of Google's and Facebook's method and practices is their invasive and non-transparent collection of users' data and its unfair and unaccountable and non-transparent use (Pasquale 2015, 55). The blurring and blending of their information practices with advertising, just as in the case of cash-for-comment and advertorials in some of the practises of the 4th and 5th Estates, becomes a much bigger and uncontrollable epistemic, ethical and eudaimonic problem when it's done at the unprecedented stratospheric global level and scale in which both Google and Facebook now operate. As Frank Pasquale poignantly puts it in regard to Google, but equally applicable to Facebook, "Don't Be Evil" is a thing of the past; you can't form a trusting relationship with a black box" (Pasquale 2015). Referring to the Myth of Gyges on which the *model of media corruption* presented in this chapter is based, Frank Pasquale correctly observes (2015,190) that,

> Black box insiders are protected as if they are wearing a Ring of Gyges – which grants its wearer invisibility but, Plato warns us in *The Republic,* is also an open invitation to bad behavior.

He goes on to ask the very crucial question that philosophers since Plato have been asking, and a question Shoshana Zuboff also asks in her book *The Age of Surveillance Capitalism* (2019) and I am now asking in this book, "what kind of society does this create?" Is it one we want and is it fit for human beings? A society in which the primary purpose of most human beings, as the Stoic and other Hellenistic philosophers and before them Plato and Aristotle clearly identified, is a good and worthy life that leads to the attainment of eudaimonia or well-being? Can we be happy in a world dominated and ruled by a handful of Tech Oligarchs, accountable only to themselves and pursuing their own self-interest over that of the common good of society? A world not of our making or design that neither is within our control, nor one that we have chosen. As Frank Pasquale astutely observes (2015),

> Accountability requires human judgement, and only humans can perform the critical function of making sure that, as our social relations become ever more automated, domination and discrimination aren't built invisibly into the [black boxes] code.

The major systemic problem identified in this chapter at the heart of Facebook's operations and analogously at the core of Google's operations, is an inherent conflict of interests between, on the one hand, Facebook's role as a media company and the normative responsibilities and commitments it has for the safeguarding of its users' information, as required by the DOIT; and on the other hand, Facebook's financial interests that are determined by an invasive and aggressive business model of maximising profits through the sale of users' data to advertisers; a business model that is by design geared to override and exert priority for Facebook's financial interests over the users' rights over their information.

Insofar as a similar systemic problem is evident in Google's operations identified also as conflict of interest between its business model that favours Google's financial interests over the normative informational rights of its users, then Google's systemic problem just as in the case of Facebook, as analysed in this chapter, is conducive to systemic media corruption. Systemic institutional corruption is, therefore, particularly problematic and harmful to society overall. That problem is analogous and by extension applicable to the Stoic problem of control applied to the problem of control of technology we examined in chapter 4. Namely, who is in control of our autonomous purposive agency that is an underlying necessary condition for our well-being? Us or the Big Tech companies, through the use of their opaque persuasive algorithms that control most if not all aspects of our lives? If as argued in chapter 4, it is the former, then we should protect and safeguard our autonomous purposive agency as both our individual and collective well-being is dependent upon it.

In a connected digital age, Stoic philosophy as we saw in chapter 4, offers a systemic unified practical philosophy of human flourishing and well-being, that combines epistemic, ethical and eudaimonic principles clustered around the central concept of wisdom, that is relevant and able to respond and motivate a theoretical and practical solution to the control problem of technology.

In keeping with the Stoic principles of living virtuously, rationally, in accordance with nature, for our individual and collective well-being, as cosmopolitan citizens with a comic consciousness for the attainment of well-being, we have, moreover, a collective responsibility, to respond to the problem of Big Tech's monopolistic control over the digital environment of the Internet, which has an overall dis-demonic impact over our lives. That solution would require a multidisciplinary and collective societal approach, which we will examine in more detail in chapters 7 and 8.

NOTES

1. Traditionally, the four estates were understood to include, the government, the clergy, the public and the press in the form of the legacy media whose role,

traditionally designated as the 4th Estate, was to hold the other three estates to account for the public good. In the advent of the Internet and the World Wide Web, where anyone with access to the Internet can disseminate and communicate information to the public, nationally and internationally, the 5th Estate is a more recent addition that includes the social media, mediated digitally via Twitter, Facebook, Google and other information technologies. The relevance and significance of the 4th and 5th Estates especially through their symbiotic convergence i.e., the *Convergent Media* (Elliott and Spence 2018, Chapter 4) is that they act as a monitoring and accountability measure to hold the other two governing estates, that of government and clergy to account. For example, it has been the Convergent Media, including WikiLeaks, that has exposed the misdemeanours and corruption of governments and clergy (police corruption and corruption of the clergy in the form of sexual abuse are two out of many examples).

2. See Spence, Edward (2017; 2008; 2006a).

3. See, for example, the Media Entertainment and Arts Alliance (MEAA) Code of Ethics in Australia, which is representative of other international code of ethics worldwide.

4. Insofar as Google, like Facebook, operates functionally as a media company in its information practices through its *search* function and its other communication and curation practices such of those of YouTube that involve the collection, profiling and harvesting of users information, which like Facebook, sells to advertisers for profit, then Google should also be viewed as being essentially a media company.

5. The 6th Estate is a neologism I introduced in this book so as to distinguish the media role of Tech companies such as Facebook and Google from that of the 4th and 5th Estates.

6. For details of that case see Cambridge_Analytica. https://en.wikipedia.org/wiki/Cambridge_Analytica (Wikipedia. Accessed: September 19, 2019).

7. By "instrumentally desirable," we mean the kind of practical prudence attributable to an instrumentally rational agent intending to act corruptly in an environment in which corruption is either illegal, or if not illegal, at least generally considered unethical. In such an environment, it would be a desirable requirement of instrumental rational agency that the agent intending to engage in corrupt activity should take the required measures to keep the agency of his corrupt actions invisible or concealed. Not doing so, could prove self-defeating and therefore instrumentally irrational. An underlined presupposition in our argument is that the agent addressed throughout this chapter is an instrumentally rational agent.

8. The *self-regarding gain* need not be merely a *self-directed* gain accruing to the corrupt agent himself. Thus, the perceived gain from the corrupt activities of some of Nixon's associates in the Watergate affair was self-regarding, in the sense that it benefited the Republican Party of which they were members, but not self-directed in the sense that it benefited or was intended to benefit those associates personally.

9. See the Shorter Oxford Dictionary.

10. For further analysis of what constitutes a conflict of interest, see Spence, E (2020) and Davies, M. (1998, 590).

11. See, for example, Zeynep Tufekci 2018. You Tube, The Great Radicalizer, The *New York Times*, March 10, 2018; Christopher Mims. 2018. Who Has More of Your Personal Data than Facebook, *The Australian and Wall St Journal*, April 23, 2018; Frank Pasquale.2015. *The Black Box Society: The Secret Algorithms That Control Money and Information*, Cambridge, Massachusetts: Harvard University Press, pp. 50–51;66;78–79;96;185;190–191. 214–215; and for a more comprehensive account of the problem, see, Shoshana Zuboff. 2019. *The Age of Surveillance Capitalism: The Fight for a Human Future at the New Frontier of Power*, Profile Books; Webb, Amy. 2019. *The Big Nine: How the Tech Titans and Their Thinking Machines Could Warp Humanity*, Hatchet Books; Galloway, Scott. 2017. *The Four: The Hidden DNA of Amazon, Apple, Facebook and Google*. London, Great Brittain: Bantam Press; Van Dijck, Poell and De Waal. 2018. *The Platform Society: Public Values in a Connected World*. Oxford: Oxford University Press; and Lynch, Michael P. 2016. The Internet of Us: Knowing and Understanding Less in the Age of Big Data, Penguin Random House.

12. The Social Contract referred to here is based on the philosophical ethical theories of Thomas Hobbes, John Locke and John Rawls that will be examined in more details in chapter 7. Although different in their approach and articulation, those social contract theories counsel every citizen and corporations to act for the common good of society as whole and not for their own exclusive self-interests in violation of the common good.

REFERENCES

Davies, Michael. 1998. *Encyclopedia of Applied Ethics.* Academic Press, Volume 1, A-D, page 590.

Elliott, Deni and Spence, Edward. H. 2018. *Ethics for a Digital Era*. Oxford: Wiley-Blackwell.

Galloway, Scott. 2017. *The Four: The Hidden DNA of Amazon, Apple, Facebook and Google*. London, Great Britain: Bantam Press.

Gewirth, Alan. 1978. *Reason and Morality*. Chicago: University of Chicago Press.

Lynch, Michael P. 2016. *The Internet of Us: Knowing and Understanding Less in the Age of Big* Data. Manhattan, New York: Penguin Random House.

Miller, Seumas. 2018. *Corruption. Stanford Encyclopedia of Philosophy*. Edited by Edward N. Zalta (Winter 2018 Edition). https://plato,Stanford.edu/archives/win2 018/entries/corruption/.

Miller, Seumas, Roberts, Peter, and Spence, Edward H. 2005. *Corruption and Anticorruption: An Applied Philosophical Approach*. Upper Saddle River, NJ: Pearson/ Prentice Hall.

Mims, Christopher. 2018. Who has more of your personal data than Facebook? *The Australian and Wall St Journal*, April 23, 2018.

Pasquale, Frank. 2015. *The Black Box Society: The Secret Algorithms That Control Money and Information*. Cambridge, MA: Harvard University Press.

Plato. 1952. *The Dialogues of Plato*. Translated by Benjamin Jowett. Chicago: Encyclopedia Britannica Inc., Book II: 359–361.

Spence, Edward H. 2021. *Media Corruption in the Age of Information*. Switzerland, AG: Springer Nature.
Spence, Edward H. 2020. The sixth estate; tech media corruption in the age of information, *Journal of Information, Communication and Ethics in Society*, 18(4): 553–573. DOI 10.1108/JICES-02-2020-0014.
Spence, Edward H. 2018. Harm in Media Marketing: The Branding of Values, in Patrick Plaisance (ed.), *The Handbook of Media Ethics*. Berlin: Walter de Gruyter GmbH, 237–251.
Spence, Edward H. 2017. Corruption in the Media, in Michael S. Aßländer and Sarah Hudson (eds.), *The Handbook of Business and Corruption: Cross-Sectoral Experiences*. Bingley, UK:, Emerald Group Publishing, 453–480.
Spence, Edward H. 2014. The Advertising of Happiness and the Branding of Values. In Michael Boylan (Ed.). *Business Ethics (2nd edition)*. Upper Saddle River, NJ: Pearson/Prentice Hall.
Snider, Mike and Baig, Edward. C. 2019. Facebook fined $5billion by FTC, must update and adopt new privacy, security measures. *USA Today*, July 24, 2019.
Spence, Edward H. Alexandra, Andrew, Quinn, Aaron, and Dunn, Anne. 2011. *Media, Markets and Morals*. Oxford: Wiley-Blackwell.
Spence, Edward H. 2011. Information, Knowledge and Wisdom: Groundwork for the Normative Evaluation of Digital Information and Its Relation to the Good Life, *Ethics and Information Technology*, Volume 13, Number, 3, 261–275.
Spence Edward H. 2009. A Universal Model for the Normative Evaluation of Internet Information, *Ethics and Information Technology*, Volume, 11, Number 4, 243–253.
Spence, Edward H. 2008. Corruption in the Media. *International Journal of Applied Philosophy*, Volume 22, Number 2 (Fall), 231–241.
Spence, Edward H. 2006. *Ethics Within Reason: A Neo-Gewithian Approach*. Maryland: Lexington Books (a division of Rowman and Littlefield, USA).
Spence, Edward H. 2006a. "Corruption doomed to fail in long term." *Financial Review*, January 14, 2006.
Tufekci, Zeynep. 2018. Facebook's Surveillance Machine, *The New York Times*, March 19, 2018.
Tufekci, Zeynep. 2018. You Tube, The Great Radicalizer, The *New York Times*, March 10, 2018.
Van Dijck, Poell and De Waal. 2018. *The Platform Society: Public Values in a Connected World*. Oxford: Oxford University Press.
Webb, Amy. 2019. *The Big Nine: How the Tech Titans and Their Thinking Machines Could Warp Humanity*. New York: Hatchet Books.
Zuboff, Shoshana. 2019. *The Age of Surveillance Capitalism: The Fights for A Human Future At The New Frontier of Power*. London: Profile Books Ltd.
Zuckerberg, Mark. 2019. Four Ideas to Regulate the Internet, *Facebook*, March 30, 2019.
Zuckerberg, Mark. 2020. Big Tech Needs More Regulation, *Financial Times*, February 18, 2020.

Chapter 7

The Normative Impact of ICT Technologies on Well-Being

Every human being is possessed of an innate human dignity that transcends all other considerations.

—Scott Morrison, Prime Minister of Australia (*The Australian*, April 30, 2021)

INTRODUCTION

This chapter will examine and evaluate the normative impact of the information and communication technologies as those used by the Big Tech companies, such as Facebook and Google, on well-being through the application of Stoic and neo-Stoic philosophy.

As we saw in chapter 6, one of the major and central topics of concern that includes normative issues such as privacy, autonomy, transparency, accuracy, accountability, truth and trust, and the illustration and contextualisation of those issues through the evaluative analysis of the Cambridge Analytica case, is *Tech Media Corruption*.

In summary, to remind us what those issues are, the object of chapter 6 was to demonstrate how some of the information and communication surveillance practices of Facebook and by analogy, Google, constitute systemic media corruption (Spence 2020). It shows that the role of Facebook, of curating and mediating the information and communication provided by its users that affords them the collective sharing and exchange of information on its platform, reveals that Facebook is essentially a *media company*, and not merely a "platform." As such, Facebook is liable to the same normative responsibilities of other media companies, based on the normative analysis of the Dual Obligation Information Theory-Wisdom (DOIT-Wisdom) model we examined in chapter 5. According to that theory, it commits Facebook, and by analogy other Big Tech companies such as Google, to a *normative editing responsibility*, qua media company, to ensure both the preventive detection and corrective editing of fake news, deep fakes, as well as other forms of misinformation and disinformation disseminated on its platforms. Based on DOIT-Wisdom's normative requirements, Facebook and by extended analogy, other Big Tech companies such as Google, is furthermore normatively accountable for the harvesting of users' information, which it collects with the aid of its persuasive algorithms and sells to advertisers and other information brokers for profit, without the users' informed consent or knowledge.

This general practice, as we saw in chapter 6, involves a *conflict of interest* between its role as a media company, and its selling of users' information to advertisers and other information brokers for profit. That conflict of interest lies at the heart of Facebook's operations as a result of its Business Model and is thus conducive to *systemic media corruption*.

Finally, as we saw in chapter 6, the systemic problem of media corruption illustrated by the practices of Big Tech companies such as Facebook and Google, is analogous and by extension applicable to the *problem of control of technology*, which we examined in chapter 4 through the theoretical lens of Stoic philosophy. Namely, who is in control of our autonomous purposive agency that is an underlying necessary condition for our decision-making choices and actions, and therefore our individual and collective well-being? Is it us, and if not, should it be us, qua citizens or the Big Tech companies, through their use of opaque persuasive algorithms that exercise informational control over most all aspects of our lives, down to our presumed purposive autonomous agency, which forms the foundation of those choices and actions? If, as argued in chapter 4, it is the former, namely, that we, as both individuals and collectively as society should be in control of our own purposive autonomous agency, and not the Big Tech companies, then we should affirm and protect our purposive autonomous agency, as both our individual and collective well-being, as citizens, is dependent upon it.

Using that Stoic principle, namely, that our individual autonomous agency should always be within our control and not the control of others as our departure point, this chapter will examine in more theoretical analytical detail how that principle applies more specifically to the control problem of technology, as it constitutes the fundamental underlying basis of that problem, fuelled by the voracious Business Models of the Big Tech companies, and in particular those of Facebook and Google.

AUTONOMY, INTEGRITY AND DIGNITY

In this section, we will examine and analyse more closely the Stoic notion of control and how it applies to technology, in terms of the concepts of autonomy, integrity and dignity that are present and implied by the Stoics' understanding of what it means to be in *control of oneself*, and in particular one's character, which according to the Stoics is essentially the only thing that is within our control and the only thing that eudaimonically matters as it relates directly to our well-being. According to the Stoics, our rational faculty comprises three distinct operations, those of *judgement, assent*, and *impulse*, and all three form part of our *prohairesis* or volition, which A.A. Long defines following Aristotle, as "choice through decision" (Long 2002, 213).

According to Long (2002, 207), *prohairesis*, "is Epictetus' favourite name for the purposive and self-conscious centre of a person." Note the similarity of Long's definition of Epictetus' term *prohairesis*, with that of the autonomous purposive agency at the centre of Alan Gewirth's argument for the principle of generic consistency (PGC) we examined in chapter 3. Recall that the PGC, as well as Gewirth's theory of *self-fulfillment* form an integral part of my Gewirthian neo-Stoic theory and its comparison with that of traditional Stoic philosophy, examined in that chapter. An equivalent term to Epictetus' preference for *prohairesis*, or volition, is *hegemonikon*, a term used by the earlier Stoics that literally means, "governing faculty." According to Long (2002, 211), "the human *hegemonikon* is the seat of rationality and the centre of the person; and in virtue of being these things it is one's epistemic, and moral disposition."

Long goes on to say that Epictetus, also uses that term in identical ways in which he uses the term *prohairesis*, as "the need to secure its 'natural' condition and also its not being controllable by another" [Epictetus Discourses, 4.5.4]. Long further states (2002, 212) that,

> Epictetus associates with *prohairesis*, his cardinal rule – the requirement to make *correct use* of our impressions. How we use them is up to us, because it falls within the purely internal domain of judgement, assent, and impulse . . . Epictetus chose the term *prohairesis* to pick out the human mind in this more restricted aspect: "us," so to say, in just those respects that are dependent on nothing that we cannot immediately judge, decide, and will "entirely *by and for ourselves* [emphasis added] . . . Epictetus associates *prohairesis* with 'the things up to us'," as distinct from external things; and he often calls these two types of thing *prohairetika* and *aprohaireta,* with the latter term signifying that everything outside the scope of *prohairesis*, including our bodies, is "not up to us."

Along similar lines as Long, Pierre Hadot quoting Epictetus, states (1998, 83),

> What depends on us are value-judgements (*hypolepseis*), impulses towards action (*horme*), and desire (*orexis*) or aversion; in a word, everything which is our own business. What does not depend on us are the body, wealth, honors, and high positions in office; in a word, everything which is not our own business. (Epictetus, Manual, I, 1)

He goes on to say (1998, 84),

> Here, we can glimpse one of the Stoics' most fundamental attitudes: the determination of our own sphere of liberty as an *impregnable islet of autonomy*,[emphasis added] in the midst of the vast river of events and or Destiny. What depends on us are thus the acts of our soul because we can freely choose them. We can judge or not judge, or not judge in whatever manner we please; we can desire or not desire; will or not will. By contrast, that which does not depend on us – Epictetus lists our body, honors, riches, and high positions of authority – is everything that depends upon the general course of nature. Our body, first it is true that we can move it, but we are not completely in control of it . . . Thus, the Stoic delimits a centre of autonomy – the soul, as opposed to the body; and a guiding principle *hegemonikon* as *prohairesis*, to the rest of the soul. It is within this guiding principle that freedom and our true self are located. It is also there, and only there, that moral good and evil can be found, for the only moral good and evil are voluntary good and evil . . . Here we have the central node of the whole of Stoicism: that of inner discourse . . . As Epictetus and Marcus Aurelius never tire of saying, everything is a matter of judgement. It is not things themselves that trouble us, but our representations of these things, the ideas we form of them, and the inner discourse which we formulate about them.

I quoted Long's and Hadot's passages above in full, because together they capture the essence of the fundamental principles of freedom and autonomy in Stoic philosophy, as well as the related concepts of *integrity* and *dignity* to be examined for further elucidation below, though their intrinsic relation to the concept of autonomy. Gewirth, as we saw in chapter 3, bases his moral philosophy, and derives his universal natural rights to freedom and well-being, on the basis of rational purposive agency. Although those rights are prima facie and apply to individuals by virtue of being prospective purposive agent (PPAs), those rights (rights to freedom and well-being) apply to *persons*, absolutely. This *dual standpoint* and perspective of a human being as at once an *agent* and a *person* is fundamental to Gewirth's moral philosophy as it is to that of Immanuel Kant's, for which the formulation of his *Categorical*

Imperative, especially in the form from which his argument of *The Kingdom of Ends* derives, is that human beings should always be treated as ends in themselves worthy of inalienable respect or dignity and never as mere means for the ends of other people.

From what has been said so far, we can see that the surveillance practices of Big Tech companies such as Facebook and Google that use persuasive technologies through the application of opaque and inscrutable algorithms to target and harvest users' information for profit, without their knowledge and consent, violates the fundamental rights of users qua persons, to be treated as ends in themselves worthy of respect and not merely as means for the financial benefit of Big Tech companies. From a Stoic and neo-Stoic perspective, it violates a person's fundamental autonomy and well-being as outlined above. It is as *persons* wherein our fundamental and absolute natural right to our autonomy and that of our dignity, inheres.

It is, as Pierre Hadot, eloquently describes it, "our own sphere of liberty as an impregnable islet of autonomy, in the midst of the vast river of events and or Destiny" and where "the Stoic delimits a center of autonomy." And importantly for our discussion in this chapter, Hadot goes on to point out that, "it is within this guiding principle [the *hegemonikon* as *prohairesis*] that freedom and our true self are located . . . Here we have the central node of the whole of Stoicism: that of inner discourse." Appropriately, the title of Hadot's book from which the aforementioned quoted passages are cited, is *The Inner Citadel* (1998).

THE CONCEPT OF SELF-RESPECT

Given the central importance to Stoic and neo-Stoic philosophy and particularly with regards to the parallel control problem for Stoicism and technology we examined in chapter 4, this section of the chapter will examine in greater detail the concept of self-respect or dignity and its intrinsic relationship to autonomy and integrity. For the purpose of this chapter and generally throughout this book, I will, unless otherwise stated, use the concepts of self-respect and dignity interchangeably.

Self-respect as a Primary Good: A Question of Life and Death

Rawls states that "self-respect and a sure confidence in the sense of one's own worth is perhaps the most important primary good" (1972, 396). He also states that "it is clearly rational for men to secure their self-respect" (Rawls 1972, 178). Gewirth's argument for the PGC demonstrates the truth of both these statements by showing how one's self-respect emanates from one's own rationality.[1] Take, for example, the case of someone, let's call him Arthur, who is being diagnosed with Alzheimer's decease and he is being told on good medical evidence that he will soon lose his memory and identity as a person, and consequently the source of his self-respect through his own sense of freedom and well-being as a person.[2] It is primarily because his well-being is diminished to such an extent that Arthur can no longer live his life with dignity. He also recognises that because he is free to act, he ought to safeguard his dignity by ending his life before the onset of his Alzheimer's condition. And he recognises that if he were impeded from carrying out his plan to end his life, he would face a worse indignity than the one he is currently facing. He would be at the mercy of other people's choices, but would have no mercy from his suffering. Thus we can see that Arthur's freedom and well-being are essential to his self-respect; his well-being because the continuous harm to it caused by his debilitating illness diminishes day-by-day his dignity and self-respect, his freedom because without it he cannot act to safeguard and retain his self-respect. Therefore, to preserve his self-respect, Arthur must have both his freedom and well-being. His freedom and well-being are not just necessary for his accomplishing any of his chosen purposes, they are also necessary for the preservation and maintenance of his self-respect and his sense of worth as a human being.

In this limiting case, we can see that freedom and well-being are not only necessary for accomplishing any of one's chosen purposes, but they are also, and perhaps more importantly, necessary for preserving and maintaining one's self-respect. This is an important discovery. First, because common experience tells us that a life without self-respect is not worth living. By "self-respect" I do not mean some inflated sense of the term, but only some minimal self-respect which allows a human being to function minimally as a normal person; that is, having enough freedom and well-being to live one's own life. The call used by the Greek partisans in their war of independence against the Ottomans which began in 1821 after continuous Ottoman occupation for four hundred years, was "Freedom or Death."[3] It signified that the Greek people could no longer live a life that was not worth living, a dignified life, under conditions approximating that of slavery. Freedom for them was a minimal essential condition for their sense of self-respect and dignity, and importantly their identity, and hence essential for a life worth living. Thus, self-respect secured by one's freedom and well-being is essential for a worthy life simply because it is one's self-respect which endows one's life with worth and value.

Second, even if the physical necessities conspire to render one's life unlivable, as in Arthur's case, so that one can no longer function even minimally as a normal person – that is, having enough freedom and well-being to pursue one's chosen purposes unimpeded by any adverse external conditions that render one's life impossible to live with dignity – one can still escape from those necessities by taking one's own life. It is, as it were, the final triumph of the will over fortune. That is not to say that one cannot triumph over misfortune by Stoically suffering "the slings and arrows of outrageous fortune." However, it requires exemplary courage and forbearance to continue living under unbearable conditions due perhaps to some sense of duty. But such saintly forbearance would, under normal circumstances be seen as supererogatory and not possible for normal, less saintly persons. Perhaps this is the question that Hamlet intends in the opening sentence of his famous soliloquy: "To be or not to be-that is the question." Albert Camus in his book *The Myth of Sisyphus*, thought that this was *the* fundamental question of philosophy:

> There is but one truly serious philosophical problem and that is suicide. Judging whether life is or is not worth living amounts to answering the fundamental question of philosophy. All the rest-whether or not the world has three dimensions, whether the mind has nine or twelve categories – comes afterwards. These are games; one must first answer. (Camus 1975, 11)

This is also perhaps Socrates' question, with which Bernard Williams begins his remarkable book *Ethics and the Limits of Philosophy:* "It is not a trivial question, Socrates said: what we are talking about is how one should live" (Williams 1985, 1). It is indeed not a trivial question, and I am inclined to agree with Camus that it is in fact the most fundamental question of philosophy. For unless we can answer that question – answer it for ourselves – life will be merely a series of meaningless "games" without any real significance. Such a life would be empty and devoid of meaning. It is perhaps this possibility that Socrates had in mind when he said that "the unexamined life is not worth living." It is by constantly examining one's life that one can best determine what a good life is for *one* to live. Which is, as we saw above, the question that Stoics were asking. But whatever life one chooses to live, given one's own particular personal characteristics and circumstances, that life will only be worth living if one can live one's life with at least a minimal degree of self-respect. It is a common phenomenon that when one's self-respect is diminished, one's respect for others and for society generally is also diminished. This is why self-respect is of paramount importance to morality. It is, I will suggest, the cornerstone of morality. As Aristotle tells us, "all friendly feelings for others are expressions of a man's feelings for himself" (Aristotle, 1976, 300). It is thus important to encourage the kind of beliefs and establish the kind of institutions in our society, both at the social and political levels, which foster a sense of self-respect in each one of its citizens. Respect is a socially symmetrical relation: self-respect promotes respect for others, and others' respect for oneself promotes and helps maintain one's self-respect. One could say that a morally stable society is a "mutual adoration" society. We shall later explore more fully in this chapter the relation between self-respect and respect for others. Here, I merely want to emphasise the importance of self-respect to morality and how Gewirth's argument demonstrates that importance.

In sum, Gewirth's argument demonstrates the importance of self-respect for morality by first showing how freedom and well-being, as the necessary conditions of action, are essential for one's self-respect. Not only is one unable to act without one's freedom and well-being, but one is unable to live one's life without a certain minimal degree of dignity and self-respect. In the aforementioned example of Arthur's decision to exercise his freedom and end his life because he no longer wishes to live without dignity is a case in point. In Arthur's case, his freedom has a poignant significance which echoes Socrates' question "How should one live one's life?" and Hamlet's similar question "To be or not to be?" In choosing to end his life, Arthur must of necessity be free to do so. By exercising his freedom for the last time, Arthur chooses to escape the very constraints of physical necessity which render his freedom necessary. In choosing to end his life, Arthur exercises his freedom *to end* his freedom. Thus, by choosing to escape through death the physical necessities which render, in his circumstances, his life unbearable and not worth living, Arthur exercises the most radical freedom of all – the freedom of choosing to cease to exist.[4]

Perhaps this is one way of understanding Kant's notion of *transcendental freedom*, although I do not wish to press this point too far, for I want to avoid engaging in any kind of metaphysical speculation. Nevertheless, it does seem that in choosing to end his life and escape the physical necessities which render his life, in his present circumstances, unlivable, Arthur is in a sense exercising a transcendental kind of freedom, transcendental because at least in its intention, it carries with it the power to release him from the physical necessities that render his life, from his point of view, not worth living. It is "freedom" understood in this radical sense that illustrates why one's freedom, at least in the extreme case we have been examining so far, carries with it a necessity which renders it, for oneself as a purposive agent, simply more than just desirable. It is, I contend, this radical necessity, as understood by an agent in the limiting case when faced with Hamlet's question of whether or not one's life is worth living, that renders one's freedom and well-being not just desirable, as a holiday or a new car, but as things whose possession is, because they are essential to one's self-respect, one's right.

Following on from our discussion above, I want to suggest that Gewirth's agent values his freedom and well-being not merely instrumentally and conatively, as necessary means for accomplishing all his chosen purposes, but as the essential components of his own sense of integrity and self-respect. Williams states (Williams, 59), "good for me introduces some reference to my interests or well-being that goes beyond my immediate purposes, and my freedom is one of my fundamental interests." For Gewirth's agent, his freedom and well-being are two of his fundamental interests because they contribute directly and essentially to his own sense of self-respect, which is an essential feature of what it is to be a person. In recognising that my freedom and well-being are essential components or characteristics of my self-respect, and not merely necessary instrumental means of carrying out any of my purposive actions, I am, as Williams poignantly puts it, beginning "to touch on some deeper questions about my conception of my own existence" (Williams 1985, 59).

Freedom and Well-Being the Fundamental Constituents of Self-respect

Gewirth's argument for the PGC reveals that a person has rights to his freedom and well-being in virtue of being a PPA, a prospective purposive agent. The analysis of the concept of "self-respect" reveals that a person needs to have the property or quality of self-respect to function fully as a person. But to have the property or quality of self-respect, which is essential and fundamental to being a person, an agent must have freedom and well-being, since freedom and well-being are the essential and fundamental constituents of a person's self-respect. Thus an agent must not only claim rights to his freedom and well-being on the basis that these are the necessary conditions for all his purposive actions, but he must also claim rights to his freedom and well-being because these are the essential and fundamental constituents of his self-respect. In sum, an agent must consider that he has rights to his freedom and well-being not only because he is the sort of being who engages in voluntary and purposive action-that is to say, a being who is a *PPA* – but also because he is the kind of being who needs self-respect – that is to say, a being who is a *person*. To be sure, by being a person an agent is also a PPA. However, our analysis above is meant to highlight what I consider to be another important and fundamental aspect of being a person, apart from merely being a PPA who engages in voluntary and purposive action.

We can now see that a person has the generic rights to freedom and well-being in virtue of being a person irrespective of what he does or omits to do as an agent. For every person, no matter what he does or fails to do, needs his self-respect. Because all persons need their self-respect equally in virtue of being persons, each person will need a certain degree of freedom and well-being, especially the latter, to preserve and maintain a minimal degree of self-respect so as to preserve and maintain his personhood. To the extent that a person has a right to have enough freedom and well-being to maintain his self-respect, that right is absolute. The right to minimal freedom and well-being, sufficient for a person to preserve and maintain his self-respect, cannot be removed without at the same time removing the very conditions necessary for an agent's personhood. Gewirth himself argues for absolute rights. According to Gewirth (1982, 219),

> A right is *absolute* when it cannot be overridden in any circumstances, so that it can never be justifiably infringed, and it must be fulfilled without any exceptions.
>
> The idea of an absolute right is thus *double normative* [emphasis added]; it includes not only the idea, common to all claim rights, of a justified claim or entitlement to the performance or non-performance of certain actions, but also the idea of the *exceptionless justifiability* [emphasis added] of performing or not performing those actions as required.

THE AGENT'S DUAL STANDPOINT

Gawith's idea of the double normativity of absolute rights Kantian in spirit, and accords with what Kant himself says about personhood in the *Grounding for the Metaphysics of Morals* (Kant 1981, 36),

> Rational beings are called *persons* inasmuch as their nature already marks them as ends in themselves, i.e., as something which is not to be used merely as means and hence there is imposed thereby *a limit* on all arbitrary use of such beings, which arc thus *objects of respect*. Persons are, therefore, not merely subjective ends, whose existence as an effect of our actions has a value for us; but such beings are *objective ends,* i.e., *exist as ends in themselves* [emphases added].

Gewirth claims, as we saw earlier, that treating persons as ends in themselves emanates directly from the PGC. In the following illuminating passage, Gewirth claims that treating persons as ends by respecting their rights to freedom and well-being provides a Kantian answer to the substantive question of moral philosophy:

"Of which interests of other persons ought one to take favorable account?"

Gewirth's Kantian answer to this substantive question is, "that the fundamental interests in question, deriving from the necessary content of action . . . are freedom and wellbeing, which are, respectively, the procedural and the substantive necessary conditions and generic features of action." Gewirth goes on to say that "to treat persons as ends in themselves is to respect their needs for the necessary conditions of action, by not interfering with them and, in certain circumstances, by helping persons to have or maintain them." (Gewirth, 1991, 92)

This way of understanding Gewirth's argument, as one that requires and agent to regard and value his freedom and well-being both instrumentally and constitutively, or intrinsically, accords well with Kant's claim (Kant 1981, 53) that an agent or a person,

> Has *two standpoints* [emphasis added] from which he can regard himself and know laws of the use of his powers and hence of all his actions: first, insofar as he belongs to the world of sense subject to laws of nature (heteronomy); secondly, insofar as he belongs to the intelligible world subject to laws which, independent of nature, are not empirical but are founded only on reason.

However, unlike Kant's "two standpoints," which are two separate worlds, the empirical spatial-temporal and the intelligible or noumenal world, the two standpoints I have been attributing to an agent above, in accordance with reconstruction of Gewirth's argument, belong to the same, one natural world. It is, I believe, one of the great advantages of Gewirth's argument, based on my reconstruction, that it can account for and explain both those two different but aligned dual standpoints of an agent, from within the natural world without having to resort to Kant's problematic metaphysics involving two distinct and ontologically independent worlds, the natural and the noumenal.

Whereas as Kant and previously Plato supported a dualistic metaphysical perspective comprising a natural and intelligible world, Gewirth as well as the Stoics as we have seen previously (chapters 3 and 4) attribute the dual standpoint of an agent within the one natural world. This is also in keeping with my argument for a neo-Stoic philosophy attributed to Gewirth, in chapter 3 of this book. Analogously, Hadot makes a similar distinction between the Kantian and Stoic ontological perspective with regard to an agent's dual standpoint. Quoting directly from the last pages of Kant's *Critique of Practical Reason* he states that,

> Two things fill the soul with ever-new and ever-growing admiration and awe, the more frequently and constantly one applies one's reflection to them: *the starry sky above me and the moral law within me.* These are two things which I have neither to search for, nor simply to presuppose, as if they were shrouded in darkness or plunged within a transcendent region, beyond my horizon: I can see them in front of me and I attached them immediately to the consciousness of my existence. The former begins at that place *which I occupy within the sensible world,* and extends my connection to that which is immensely large, with its worlds upon worlds and its systems of systems, in addition to the unlimited

times of their periodic movement, their beginning and their duration. The latter begins at my invisible self, or *personality*, and it represents me within a world which possesses a genuine infinity, but which can be detected only by the understanding, and with which (and thereby also with all these visible worlds) I realize that I am in a relationship of . . . universal and necessary linkage. The first spectacle, that of an innumerable multitude of worlds, somehow *annihilates* my importance *qua* that a bestial creature which must return to the planet – a mere point in the universe – the matter out of which it was formed, after having been – one know not how – provided with a vital force for brief span of time. The second spectacle, by contrast, *increases* my value infinitely, *qua* that of an intelligence, thanks to my personality within which the moral law displays to me a life independent with regard to animality, and even with regard to the entire sensible world. (Hadot 1998, 182, from a quotation of Immanuel Kant's, Critique of Practical Reason, Hamburg, 10th edition, 1990,186)

I have allowed myself the luxury of some space to quote this profoundly poetic passage by Kant, quoted by Hadot, in full, to display, as Hadot himself probably intended, the close affinity and relationship of Kant's dual standpoint of an agent to that of the Stoics and analogously to that of Gewirth's. As Hadot correctly observes, however, the Stoics would not have accepted Kant's distinction between a sensible and intelligent world and neither would Gewirth as we saw earlier in my reconstruction of his argument of the PGC around the concept of dignity, which also aligns the neo-Stoicism I am attributing to him, with the Stoics single natural perspective of the universe. However, for the Stoics as Hadot points out just as for Kant, though with the qualification, that Kant is referring to the intelligible world (Hadot, 1998, 182), "it is the self's awareness of itself which transforms it, making it pass in succession from the domain of necessity to the domain of freedom, and from the domain of freedom to the domain of morality. The self- that infinitesimal point within the immensity – is thereby transformed, and made equal to universal Reason."

The transformation to which Hadot refers in the passage above should remind us of the Stoic concept of *oikeiosis* we examined in chapter 3. That is, by one's character being transformed by virtue and rational action, in agreement with nature, one becomes part of universal reason that according to the Stoics permeates the whole of the universe and by virtue of which human beings are related by their common rationality they share with each other and the whole natural world, or Cosmos. Thus, our natural universal reason bequeaths to us our inviolable dignity as persons and unites us as citizens of a cosmopolitan community, to which, though not directly by reference to the Stoic view, Gewirth, appropriately refers to as a *Community of Rights*, which is the title of his 1996 book.

AUTONOMY AND INTEGRITY

According to A. A. Long, Epictetus connects autonomy with the "correct use of impressions." He goes on to say that, "impressions, are causal only in the sense that they make us aware of their objects" but it is entirely up to us what we do with them, as it is up to us whether we ascent to them or reject them. If they make us ascent to something bad that goes against our moral character the consequence of such ascent, according to Long (2002, 216) is that,

We surrender our own agency and put ourselves in the position of being "conquered" or "disturbed" or "dazzled" by them . . . the outcome of such ascent is an ailment of rationality or a passion . . . that along with loss of autonomy we may well allow the impression to make us pursue unethical objectives . . . and so suffer loss of integrity.

He goes on to say that "Epictetus' crucial point is that "impressions have only the effects that we permit them to have" (2002, 214–216),

> Human impressions, as conceived by Epictetus and earlier Stoics, generally have the propositional structure, which is to say that they are either true or false . . . Yet, because apparent truth is far from equivalent to actual truth, her takes the task of judging and interpreting impressions to be the critical test of human rationality, consistency, and moral character.

The connection between autonomy and integrity for the Stoics comes, as Long describes in the above passage and in many parts of Epictetus' *Discourses*, from the rational and moral consistency of an agent's actions, in which the agent's rational judgements and moral actions are integrated harmoniously and consistently within the agent's rational and virtuous character, at all times, regardless of the contingent external circumstances that for the most part are, outside the agent's control, and hence of no concern to them. According to Long that crucial claim Epictetus and generally the Stoics make regarding our *prohairesis*, which Long interprets as *volition* (2002, 220),

> Is that nothing outside our individual selves has ultimate authority over what we want or do not want. Epictetus presents this *capacity* [emphasis added] as the basis fact about the self. It is a claim concerning our power as agents, but as agents of our mind's behaviour rather than of what we can do to bring about changes in the external world . . . Epictetus locates the essence of the self in the mental processes and dispositions that necessarily precede every bodily action; for prohairesis literally means "pre-choice" or "choice before choice." Which is how the term is defined in Stoicism.

I highlighted the term *capacity* in the above passage, to emphasise its importance as regards the putative internal control we are presumed to have over our own autonomy and *prohairesis* in the Stoic schema of the unencumbered self, free of external circumstances that are not up to us, and hence not within our control. As Long correctly observes (2002, 221), "the free will that Epictetus promulgates is not a universal psychological datum but an arduous project that is equivalent to mastery of Stoic philosophy." In the following passage that Long refers to, Epictetus presents us with clear description of the *capacity for freedom* that all human agents naturally have within them:

> The unconstrained person, who has access to the things that he wants, is free. Whereas he who can be constrained or necessitated or impeded or thrown into anything against his will is a slave.
> *Who is unconstrained?*
> The person who seeks after nothing that is not his own.
> *What are the things not one's own?*
> Everything that is not up to us to have or not to have, or to have thus and so qualified, or thus and so disposed. Therefore, the body is not one's own, not its members, nor property. If, then, you attach yourself to any of these things as if it were your own, you will pay the appropriate penalty of on one who seeks after things that are not one' own. The road that leads tom freedom, and the only release from enslavement, is to be able to say wholeheartedly: Lead me, Zeus and you, Fate, wherever you have ordained for me (Epictetus 4.I. 128–31).

Commenting on the passage of Epictetus above, Long correctly observes (2002, 222) that, "to the extent that Epictetus connects freedom and with fate, he does so be treating the autonomous person as someone who voluntary complies with a predetermined situation rather than by asking how a causal nexus can be compatible with a capacity for assent that is completely unconstrained."

The importance of Long's clarification of Epictetus' passage where he connects freedom with fate cannot be emphasised enough. For one who is committed to a Stoic way of life as regards Stoic ethics, but sceptical about the deterministic metaphysical theory of the Stoics, can still act in accordance with Stoic ethics and the Stoic rationale of only being concerned of things within one's rational control, without having to subscribe to the optimistic metaphysical doctrine of the Stoics that this is the best and most rational possible world. One could in fact adopt an Epicurean metaphysical perspective of the universe, but still live one's life in accordance with the rational and ethical precepts recommended by Stoic philosophy. As I argued in chapter 3 in my comparison of a Gewirthian neo-Stoic philosophy to that of the traditional Stoic philosophy, the neo-Stoicism I proposed, similar in all other respects to Stoic philosophy, has the theoretical advantage of not being committed to any particular metaphysical view of the universe as Stoicism does. However, this only emphasises further the point I was making earlier, that commitment to Stoic metaphysics is not necessary for a commitment to a Stoic ethical way of living. This is reinforced by the view Marcus Aurelius expresses in his *Meditations,* namely, that whether one accepts the Epicurean or Stoic metaphysical view, both views are consistent with Stoic ethics (quoted by Hadot, 1998, 148):

> If the All is God, then all is well. But if it is ruled by chance, don't you, too, be ruled by chance (IX, 28, 3).
> Consider yourself fortunate if, in the midst of such a whirlwind, you possess a guiding intelligence within yourself (XII, 14, 4).
> On either hypothesis, then, we must maintain our serenity and accept events by the way they are. It would be just as crazy to blame atoms as it would be to blame the gods (VI, 24).
> Our choice of a model of the universe thus changes nothing with regard to the fundamental Stoic disposition of consent to events, which is nothing other than the discipline of desire (X, 7, 4).

Hadot summarises Marcus' position on the two metaphysical opposing views of the universe and their moral implications, as follows (1998, 148):

> Marcus thus opposes two models of the universe: that of Stoicism and that of Epicureanism. His reason for doing so is to show that, on any hypothesis, and even if one were to accept, in the field of physics, the model most diametrically opposed to that of Stoicism, the Stoic moral attitude is still the only possible one. If one accepts Stoic physical theory – that is to say, the rationality of the universe – then the Stoic moral attitude – that is, the discipline of desire, or rational consent to the events brought about by universal Reason – does not raise any difficulties: one must simply live in accordance with reason. If, however, one accepts the Epicurean physical theory – a model where the universe is a dust of atoms produced by chance and lacking unity – then the grandeur of humankind consists in our introduction of reason into the chaos.

The above passages once more emphasise the crucial part *autonomy, integrity,* and *human* dignity, and their active expression and application through *prohairesis* or *hegemonikon,* play in Stoic ethics. In *Pensées,* Blaise Pascal referring to Man as a "thinking reed" expresses the *grandeur of humankind* in the following resonant and evocative passages (Pascal, 1966, p.59):

> *Thinking reed.* It is not in space that I must seek my human dignity, but in the ordering of my thoughts. It will do me no good to own land. Through space the

universe grasps me and swallows me up like a speck; though thought I grasp it. (113:348, p.59)

Man's greatness comes from knowing he is wretched: a tree does not know it is wretched. Thus it is wretched to know that one is wretched, but there is greatness in knowing one is wretched. (114: 397, p. 59)

Man's greatness. Man's greatness is so obvious that it can even be deduced from his wretchedness, for what is nature in animals we call wretchedness in man, thus recognizing that, if his nature is today like that of animals, he must have fallen from some better state which was once his own. (117: 398, p. 59)

Similarly, Albert Camus expresses an analogous evocation to the human spirit, in his book *The Myth of Sisyphus*. In the absurdity of Sisyphus' punishment by the gods of pushing a rock up and down a hill for all eternity repeatedly, with no respite nor reprieve, he pinpoints humanity's dignity and happiness. For as he says (1955, 110–11),

One does not discover the absurd without being tempted to write a manual of happiness . . . It happens as well that the feeling of the absurd springs from happiness. "I conclude that all is well," says Oedipus, and that remark is sacred. It echoes the wild and limited universe of man. It teaches that all is not, has not been, exhausted. It drives out of this world a god who had come into it with dissatisfaction and a preference for futile suffering. It makes of fate a human matter, which must be settled among men.

All Sisyphus' silent joy is contained therein. His fate belongs to him. His rock is his thing . . . There is no sun without shadow, and it is essential to know the night . . . For the rest, *he knows himself to be the master of his days* [emphasis added] . . . The rock is still rolling.

I leave Sisyphus at the foot of the mountain! One always finds one's burden again. But Sisyphus teaches the higher fidelity that negates the gods and raises rocks. He, too concludes that all is well. This universe henceforth without a master seems to him neither sterile nor futile. Each atom of that stone, each mineral flake of that night-filled mountain, in itself forms a world. The struggle itself towards the heights is enough to fill a man's heart. *One must imagine Sisyphus happy* [emphasis added].

I have quoted Camus' passage in full above, to also emphasise two related and crucial points following our discussion so far: the first, is that regardless of the different metaphysical perspectives of Epicureanism, Stoicism and those of Pascal's Christian and Camus' Existentialist perspectives, humanity holds the centre stage, by both deep moral sentiment and rational thinking. The second point, key to all those different but related metaphysical perspectives, are freedom, integrity and human dignity, that makes us all capable masters over our own lives, as indicated by the Stoics in their systematic rational principle, of only being concerned with things within our control, such as our moral character, and indifferent to all else that lies outside our control. For then, and only then, it makes perfect sense to imagine that even Sisyphus, pushing his rock up that hill, can be happy. That is, if he thought and acted like a Stoic.

THE DIGNITY-CONFERRING VALUE OF RIGHTS

Significantly, Gewirth's PGC argument for universal natural rights to freedom and well-being – on the basis of which in chapter 3, I proposed a neo-Stoic philosophy, centred around Gewirth's theory of self-fulfillment – establishes that agents *qua* persons have not only prima facie rights as agents, but moreover, have absolute rights to their dignity or self-respect. The dignity-conferring value of absolute rights is both crucial and significant because we can now claim that not only Big Tech companies, such as Facebook and Google, undermine peoples' personal autonomy through their opaque persuasive technologies and inscrutable algorithms, but moreover, violate the absolute right people have to their dignity as persons; as well, as the rights they have to their freedom and well-being on which their dignity as persons necessarily depends. Furthermore, the erosion of peoples' dignity as individuals also undermines our collective sense of a cosmopolitan community and the moral solidarity such a community can provide. This is also a problem for our efforts in mounting collective action in requesting that Big Tech companies cease from undermining our common rights and values as users and citizens of their platforms, such as rights to our autonomy, freedom, privacy, trust and accountability, and our absolute right to our dignity as persons, which the Big Tech companies systematically violates as we saw in chapter 6, and therefore undermine individually and collectively, our well-being.

The distinction between being an agent and being a person can be clearly demonstrated in terms of the harm that a person may suffer as an *agent* and the harm he may suffer as a *person*. We can clearly conceive of a person suffering a certain harm as a result of his freedom and well-being being interfered with by others, with regard to the agent's purposive actions, with no loss of self-respect, and we can also clearly conceive of a person suffering a loss of self-respect as a result of being degraded by others in some way, without any hindrance to the performance of any of the agent's purposive actions. In the first instance, the agent would suffer, as an agent, an *instrumental* harm by virtue of not being able to perform some of his purposive actions. In the second instance, the agent would suffer, as a person, a *personal* harm by virtue of suffering a loss of self-respect.

The distinction I have been making between, on the one hand, being an agent and being a person, and on the other, suffering an instrumental harm as an agent and suffering a personal harm as a person, is given strong textual support by what Gewirth himself says in his paper "Why Rights are Indispensable":

> When A has a right to X, the protection of his interest in X is justified because he has a personal title to have X so that X is personally owed to A as his due and for his own sake, not because it adds to overall utility. If X is withheld from A, then not only is he harmed-some important interest of his is adversely affected-but he is also *personally wronged* [emphasis added], in that he is prevented from having something that belongs directly to him. (Gewirth 1986a, 335)

According to Gewirth, a crucial aspect of rights, and in particular generic rights, is that they have a *"personal orientation* which derives both from the justifying ground and from the right-holder's relation to the duty-bearer" (1986, 334–35). According to Gewirth's argument of dignity, the generic rights to freedom and well-being have "a personal orientation" precisely because they relate to an agent *personally* and *constitutively or intrinsically,* by virtue of freedom and well-being being constitutive of his self-respect, and not merely *instrumentally,* by virtue of freedom and well-being being the necessary means for the accomplishment of any of the agent's purposive actions. It is because of this personal orientation, which the generic rights have with regards to a person's self-respect, that an agent can be "personally wronged" in addition to being instrumentally wronged. And being personally wronged, the agent is "prevented from having something that belongs directly to him" – according to my reconstruction of Gewirth's argument, his *self-respect* as a person.

In "Why Rights are Indispensable," Gewirth talks of the "dignity-conferring value of rights." Gewirth identifies the source of the dignity-conferring value of rights in "the position of being a claimant that pertains to every right-holder of suitable age and mental capacity" (1986, 335–36). According to Gewirth's argument, having rights, specifically generic rights, confers dignity on persons precisely because dignity is essentially and fundamentally constituted by the objects of those rights, namely, freedom and well-being.

"The position of being a claimant" is "an important source of the dignity that having rights confers on persons" (Gewirth 1986, 335) precisely because, according to Gewirth's argument, freedom and well-being, as the generic features of agenthood and personhood, are both the objects of the generic rights which an agent claims for himself and the essential and fundamental constituents of the property of personal dignity which the agent possesses by virtue of being a person. It is because of this *intrinsic* relation between, on the one hand, freedom and well-being as the objects of the generic rights, and on the other, freedom and well-being as the essential·and fundamental constituents of the personal property of dignity, that the claiming of the generic rights by an agent has a "dignity-conferring value." That is, they have a dignity-conferring value precisely because the objects of those rights are simultaneously the essential and fundamental constituents of the property of dignity. If freedom and well-being were merely valued as the necessary instrumental *means* for action, and were not also valued for themselves as the essential and fundamental constituents of an agent's dignity, then the claiming of the generic rights by an agent would not have the dignity-conferring value that Gewirth refers to. Of course, as merely the necessary means for action, freedom and well-being will have a prudential value for the agent. But if the agent was not also a person, who valued his freedom and well-being as ends in themselves because he considered them essential for the preservation and maintenance of his dignity, the agent's rights to his freedom and well-being will lack the dignity-conferring value which his generic rights have by virtue of his being a person for whom personal dignity matters.

THE SOCIAL DIMENSION OF PERSONAL DIGNITY

The double aspect of dignity is also suggested by Gewirth in *Reason and Morality:*

> The PGC requires that this self-regard [self-esteem] include a social dimension in two directions. The agent's self-esteem must be reflected in a corresponding esteem for him on the part of other persons, and he must himself have a corresponding esteem for other persons. (1978, 242)

Although Gewirth's passage above refers to the double aspectness of dignity with regard to its *direction* rather than its *composition,* it is nevertheless clear that Gewirth recognises the *social dimension* of the concept of dignity. My own explanation of the concept of dignity ascribes to it a double aspectness both with regard to its composition and its direction. With regard to its composition, I have argued above that a person's dignity has both a personal and a communal component, comprising, respectively, the agent's freedom and well-being and those of the community to which he belongs. The mix between these two components in a person's dignity varies, of course, from person to person, depending on the degree of both his personal and his communal affiliations and commitments. In some persons one of the two components may, under certain circumstances, overrun and override the other component: in some cases the personal ill overrun and override the communal component of dignity, as for example in the case of the sociopath; in other cases, by contrast, the communal will overrun and override the personal component of dignity, as for example in the case of the communal fanatic as illustrated by suicide bombers. In both cases, the socially disorientated loner or sociopath, as well as the depersonalised fanatic, exhibit a moral disregard and disrespect for other individuals as persons that is in direct violation of the PGC's requirement of mutual and equal respect for all. This moral disregard for other persons in their double aspect, as both individuals and members of their communities, also goes directly against one of the fundamental principles of Stoic philosophy, that of cosmopolitanism, which we examined in chapters 3 and 4. Significantly, we can now see more clearly and directly, through my reconstruction of Gewirth's argument for the PGC around the crucial concept of dignity, that the social and communal dimension of dignity I have been ascribing to Gewirth in this chapter, is consistent with and of a piece with Stoic cosmopolitanism, as an expression of fundamental universal rights that people have a right to, by virtue of being *persons* worthy of inviolable respect.

By revealing the centrality of dignity in morality, Gewirth's argument is able to explain the pervasiveness and resonance of morality. It explains it by revealing that indignities suffered by one person as a result of the actions of others affect and harm not only that person and his community, but also the human race as a whole. Extreme degradations suffered by persons at the hands of others, harm and degrade both the agents and the recipients, as well as harming and degrading all of us as persons, at least symbolically, by degrading the universal minimal worth that all of us are due by right; that is, by virtue of our common humanity-by being persons worthy of respect.

CONCLUSION

Gewirth's argument for the PGC, which demonstrates that each person has a right to be treated by others respectfully, is in perfect reflective equilibrium with our shared belief that each person is worthy of respect just because he is a person, and thus should be treated by others respectfully. In addition, Gewirth's argument for the PGC around the double-aspect notion of dignity, demonstrating how each person is required by his own reason to respect not merely other persons' prima facie rights to their freedom and well-being (as the necessary means to their purposive actions as agents) but also their absolute rights to freedom and well-being (as the essential and fundamental constituents of their personal dignity and personhood), resolves Kant's apparent and metaphorical "paradox":

That merely *the dignity of humanity* [emphasis added] as rational nature without any further end or advantage to be thereby gained – and hence respect for a mere idea – should yet serve as an inflexible precept for the will; and that just this very independence of the maxims from all such maxims from all such incentives should constitute the sublimity of maxims and the worthiness of every rational subject to be a legislative member in the kingdom of ends, for otherwise he would have to be regarded as subject only to the natural law of his own needs. (Kant 1981, 43)

Gewirth's argument, with its emphasis on the double-aspect concept of dignity, resolves "Kant's paradox" because it shows how the "dignity of humanity" and our own personal dignity are closely interrelated through freedom and well-being, as both the necessary features of action and agency on the one hand, and the fundamental constituents of personal dignity and personhood on the other. According to Gewirth:

The ultimate purpose of the [generic] rights is to secure for each person a certain fundamental moral status: that of having rational autonomy and dignity in the sense of being a self-controlling, self-developing agent who can relate to other persons on a basis of mutual respect and cooperation, in contrast to being a dependent, passive recipient of the agency of others. (Gewirth 1985, 743)

The "fundamental moral status" of each person is demonstrated by Gewirth's argument simply, realistically and naturally, without any unnecessary metaphysical devices such as Kant's "noumenal self." Gewirth's argument for the PGC, by contrast, demonstrates that by virtue of freedom and well-being as the necessary conditions of agency and the fundamental constituents of personhood, persons have, categorically, a rational and moral obligation to be benevolent to each other by at least respecting each other as persons, because persons, as ends in themselves, are worthy of respect. Gewirth's argument, achieves this realistic minimal benevolence both simply and naturally by demonstrating the rational and moral implications of our personal freedom and well-being, in our dual capacity as both agents and persons, without any unnecessary metaphysical devices and unrealistic epistemological assumptions and requirements.

It is by unpacking the concept of well-being in Gewirth's argument, as Gewirth himself does through several of his writings as well as in *Reason and Morality* (1978), that the wider moral connections between the generic rights of freedom and well-being and such notions as "the self" "character" and "community" and "society" and "cosmopolitanism" through the neo-Stoic philosophy I am attributing to him, become apparent. It has been partly my aim in this chapter to lay bare the important moral connections between the generic right of freedom and well-being and the concepts listed above.

I have endeavoured to show how the generic rights, through their basis in both action and dignity, agency and personhood, can both justify and explain several of our most prevalent and pervasive moral phenomena. It is, I believe, one of the virtues of Gewirth's argument for the PGC that it can account both for the great variety of the moral phenomena and for the resonance and pervasiveness of morality throughout our lives in the most direct, simple and natural way, as well as in keeping with neo-Stoic philosophy I attributed to him in chapter 3. As Michael Boylan might say,[5] Gewirth's theory not only provides adequate justification in support of a supreme principle of morality, the PGC, it also provides an adequate *understanding* of morality. It is able to do so, through the rich empirical content that is capable of being generated on the basis of the necessary features of purposive action, freedom and well-being, and especially, on the basis of my reconstruction of the argument, around the concept of self-respect.

As I tried to show through various examples drawn from our collective culture, as human beings we consider our personal dignity, both in its personal and communal and social manifestations, to be of paramount importance to us as persons. It defines, to a very large extent, who and what we are as persons. As I argued above, it is precisely because individual persons want to preserve their personal dignity, either for themselves or on behalf of their respective communities, that they would at times risk their own lives, or in more extreme cases, but not uncommonly, even take their own lives, or choose death to slavery or dishonour to protect, preserve and maintain their own dignity and that of their communities. Although these may be considered as limiting cases, I also tried to show, through less extreme and more commonplace and ordinary examples, that persons regard their dignity with the utmost importance and seriousness.

It is because people care, at least to some minimal degree, about their dignity that they value their freedom and well-being not only as necessary means to their purposive actions, but also as ends in themselves, by virtue of these being the essential and fundamental constituents of their dignity. For without freedom, and some degree of well-being, a person cannot preserve and maintain his dignity, and thus cannot preserve and maintain his personhood. This is my minimal thesis about personal dignity and personhood, which I believe accords well with pre-theoretical common folk beliefs on the subject, and also those embedded in our cultural notions, customs and traditions, at least in the west, spanning at least the time-frame between Homer and the present. It is, as I mentioned earlier, one of the great virtues of Gewirth's argument for the PGC that it can both justify and explain our shared common cultural beliefs and traditions regarding the concepts of dignity and personhood through the concepts of freedom and well-being.

NOTES

1. For a more extensive and detailed account of Gewirth's account of self-respect, see Spence, E.H. 2006. *Ethics Within Reason: A Neo-Gewirthian Approach*, Chapter 4. 159–213.

2. There have been several feature films of late exploring that very case such as *Supernova*. April 2021. Director: Harry Macqueen; Screenplay: Harry Macqueen; Actors: Colin Firth and Stanley Tucci. *The Father*. April 2021. Director: Florian Zeller. Actors: Anthony Hopkins and Olivia Colman; and *Black Bird*. February 2021. Director: Roger Michael; Screenplay: Cristin Torpe; Actors: Susan Sarandon; Kate Winslet and Sam Neil.

3. Gewirth refers to Patrick Henry's similar phrase "Give me liberty or give me death!" as an instance of an intrapersonal conflict where one's "duty to maintain freedom takes priority over the duty to maintain life itself" (1978, 346). The important point to note here is the prevalence of this moral sentiment in Western culture. The heroic stand of the 300 Spartans against the Persians at Thermopylae is one of the earliest and most famous examples. Masada and the Alamo are two other historical examples that express the same type of moral sentiment.

4. It is the lack of this personal freedom that makes Sisyphus' punishment of rolling a rock up and down the same hill for all eternity seem so terrible to us. For, unlike Arthur, Sisyphus cannot escape from his suffering and indignity; that is, the suffering and indignity at being reduced to perform a meaningless repetitive task for all eternity. Paradoxically, the gods must allow Sisyphus to retain enough of his personal qualities, including his sense of self-respect, so that he should suffer as a direct result of his punishment. For if he were to lose all the personal qualities that make him a normal person, and become either a madman who was not aware of what he was doing, or, through his repetitious task, a mere thoughtless and emotionless automaton who performed his task without reflecting upon it, then Sisyphus would not be aware of his punishment and would thus not consciously suffer from it, as the gods intend him to. The gods were ingenious in devising a punishment that would inflict the most terrible suffering on a rational person – the punishment of performing a meaningless task for all eternity in total isolation, without the fellowship of other persons. Sisyphus' case serves to highlight the importance of freedom to human beings, not merely in their capacity as purposive agents but also as self-respecting persons.

5. Michael Boylan, "Choosing an Ethical Theory," in *Gewirth: Critical Essays,* ed. Boylan, 75–90.

REFERENCES

Aristotle. 1976. The Nicomachean Ethics. Harmondsworth, Middlesex: Penguin Books.
Boylan, Michael.1999. "Choosing an Ethical Theory." In *Gewirth: Critical Essays On Action, Rationality And Community,* ed. Boylan. Lanham, MD: Rowman and Littlefield.

Camus, Albert. 1975. *The Myth of Sisyphus*. Translated from the French by Justin O'Brien. Harmondsworth, Middlesex: Penguin Books.
Gewirth, Alan. 1996. *The Community of Rights*. Chicago: Chicago University Press.
Gewirth, Alan. 1991. "Can Any Final Ends Be Rational." *Ethics*, 102(October 1991), 66–95.
Gewirth, Alan. 1986. "Why Rights are Indispensable." *Mind* 95, 329–44.
Gewirth, Alan. 1985. "Rights and Virtues." *Review of Metaphysics*, 38, 739–62.
Gewirth, Alan. 1982. *Human Rights: Essays: Essays on Justification and Applications*. Chicago: Chicago University Press.
Gewirth, Alan. 1978. *Reason and Morality*. Chicago: University of Chicago Press,
Hadot, Pierre. 1998. *The Inner Citadel: The Meditations of Marcus Aurelius*. Cambridge, Massachusetts: Harvard University Press.
Kant Immanuel. 1981. *Grounding for the Metaphysics of Morals*. Indianapolis: Hackett Publishing Company.
Long, A. A. 2002. Epictetus: A Stoic and Socratic Guide to Life. Oxford: Clarendon Press.
Pascal, Blaise. 1966. *Pensées*. Harmondsworth, Middlesex: Penguin Books.
Rawls, John. 1972. *A Theory of Justice*. Oxford: Oxford University Press.
Sheridan, Greg. 2021. "Transcendent plea from the heart a timely reminder." *The Australian*, 1–4, Friday, April 30, 2021.
Spence, Edward H. 2020. The sixth estate; tech media corruption in the age of information, *Journal of Information, Communication and Ethics in Society,* 18(4): 553–573.
Spence, Edward H. 2017. "Corruption in the Media," *The Handbook of Business and Corruption: Cross-Sectoral Experiences*, ed. Michael S. Aßländer and Sarah Hudson. Bingley, UK: Emerald Group Publishing, 453–480.
Spence, Edward H. 2008. Corruption in the Media. *International Journal of Applied Philosophy*, 22(2), Fall: 231–241.
Spence, Edward. H. 2006. *Ethics Within Reason: A Neo-Gewirthian Approach*. Lanham, MD: Lexington Books (a division of Rowman and Littlefield, USA).
Williams, B. 1985. *Ethics and the Limits of Philosophy*. London: Fontana Paperbacks.

Chapter 8

The Normative Impact of AI Technologies on Well-Being

The ultimate form of control is individual sovereignty, understood as self-ownership, especially one's own body, choices, and data. The fight for digital sovereignty is an epochal struggle.

—Luciano Floridi, 2019.

INTRODUCTION

In chapter 7, we examined and analysed the concept of dignity that is central to both Stoic and neo-Stoic philosophy and specifically with regard to the problem of control, as expressed in Stoic philosophy and analogously as it applies to technology, and in particular, ICTs and AI Technologies. As most of the key problems in the use of those technologies are generated through the use of AI algorithms, those problems will be analysed in conjunction with the normative impact those technologies have on well-being, on individuals and society generally.

This chapter will examine and evaluate in more detail the impact of ICTs and AI technologies on well-being[1] through a close analysis of two major problems in AI machine learning and neural network algorithms: that of the *inscrutability* or the *black box* problem and the *control problem*.[2] The chapter will also suggest possible solutions to both of those problems, some of them proposed recently by AI researchers. A key concern that both those problems raise is that without transparency there is no accountability and without accountability there is no normative way to monitor AI systems for potential adverse and harmful impacts on society, let alone control them. As such, the common collective good of society and humanity generally, is at stake.

The chapter will also introduce the notion of anticipatory normative controls through design[3] to prevent or at least minimise the risk of the adverse impact of AI technologies on well-being through opacity or lack of control. As in chapter 7, the core problem of technology examined in this chapter and generally throughout this book is, who is in control of technology, Us, or the Big Tech companies?

THE CONTROL PROBLEM AND DIGITAL SOVEREIGNTY

In his perspicuous and informative article, "The Fight for Digital Sovereignty: What It Is, and Why It Matters, Especially for the EU" (Floridi 2019), Luciano Floridi describes *Sovereignty* as "a form of legitimate, controlling power" (2019, 372). In turn, he defines "control" to mean, with regard to the notion of sovereignty (2019, 371) as,

> The ability to influence something (e.g., its occurrence, creation, or destruction) and its dynamics (e.g., its behaviour, development, operations, interactions), including the ability to check and correct for any deviation from such influence. In this sense, control comes in degrees and above all can be both pooled and transferred. This is crucial since we shall see the *ultimate form of control is individual sovereignty* [emphasis added], understood as self-ownership, especially one's own body, choices, and data. The fight for digital sovereignty is an epochal struggle . . . the most visible clash is between companies and states and it is *asymmetric* [emphasis added].

I have emphasised in the above quotation from Floridi, the key terms that are central to Stoic philosophy and their relation to the control problem of technology we encountered in the previous chapters, and in particular chapter 7. As we saw in chapters 4 and 7, for the Stoics the ultimate and only control we have as human beings is the control over our judgements, choices, and impulses that collectively the Stoics identified with our individual volition or *prohairesis* or *hegemonikon*. Crucially for the Stoics, the inner control we have over our *self,* and especially with regard to our self-respect, does not include our bodies as that is contingent and lies for the most part outside our control as it is subject to outside circumstance and conditions that are not always within our control. With regard to our *individual sovereignty* and analogously, our autonomy, we saw in chapter 7 how the Big Tech companies through the *asymmetric* power of their persuasive technologies, infiltrate, influence and manipulate our autonomy and hence, undermine our individual sovereignty and consequently, our individual and collective well-being.

Moreover, as we in chapter 6, that can be conducive to systemic media corruption, as illustrated by the *Cambridge Analytica* case. Worth noting here is that the monopolistic *hegemony* exercised by the Big Tech companies, such as Facebook and Google over society and us, is what undermines crucially, the *hegemony of our individual sovereignty*, expressed by the Stoics as our individual *hegemonikon*, our own inner volition. As Floridi correctly observes this crucial fight is largely between states and companies, such as Facebook and Google. For individual citizens are transformed in the digital age under the influence of *Surveillance Capitalism* (see Zuboff. 2019) from traditional *voter-consumers* to *follower-users* (Floridi 2019, 372). As Floridi astutely observes, "our profiles are created, owned, and exploited not just by states, but also by multinationals, which, as the word indicates are globalised." He goes on to say (Floridi 2019, 372–373) that,

> The digital age is forcing us to rethink the nature of sovereignty. Modern-analogue sovereignty is still necessary but increasingly insufficient . . . Once cumulative advantages create a monopolistic regime, then there is no real competition, so no real consumer's choice, and therefore no real accountability of the companies that dominate specific markets.

The question of who is best placed to safeguard and protect the control over our individual and collective digital sovereignty against the normative abuses of the monopolistic Big Tech companies, through appropriate regulation by the democratic states, is a topic we will examine in more detail later in this chapter. Suffice to say for now, that already several legal actions against the monopolistic anti-trust violations by the Big Tech companies are under way (as of December 2020) in the USA, the EU and Australia. Recently on July 29, 2020, Jeff Bezos (Amazon), Tim Cook (Apple), Sundar Pichai (Google) and Mark Zuckerberg (Facebook) were "questioned during a hearing of the House Judiciary Subcommittee on Antitrust, Commercial and Administrative Law on 'Online Platforms and Market Power, Part 6: Examining The Dominance of Amazon, Apple, Facebook, and Google'" (Floridi 2019, 373). This development is particularly significant for our discussion concerning the control Big Tech companies exercise over technology, and especially with regard to the control they exercise over our autonomy and digital sovereignty as individuals and collectively as users.

The problem of the current monopolistic control by Big Tech over our digital data becomes precisely from a Stoic and neo-Stoic perspective a critical problem for the legitimacy or more precisely the *illegitimacy* of that control. For it affects not only what is contingently outside our control, but crucially what ought to be legitimately within our individual control and not under the control of the Big Tech companies. For as Floridi correctly points out, that is ultimately an issue of our *individual sovereignty,* that as demonstrated in chapter 7, its legitimacy is based on our *prima facie* natural rights to our freedom and well-being as *agents,* and as *absolute* inviolable rights to our dignity, as *persons.*

THE CONTROL PROBLEM AND THE NORMATIVITY OF ALGORITHMS

In the previous section, we examined how the Big Tech companies exercise control over the sovereignty over our digital data through their monopolistic practices that are mediated by their AI persuasive technologies in the form of opaque and unaccountable algorithms. In this section, we shall examinee in more detail how and in what ways algorithms are used to undermine our normative rights as digital-agents and users. As we have seen earlier throughout this book so far and in particular in chapter 5 through the application of the Dual Obligation Information-Wisdom (DOIT-Wisdom) model, those rights comprise epistemic, ethical and eudaimonic rights.

In a ground-breaking paper, "The ethics of algorithms: Mapping the debate" (Mittelstadt et al. 2016), the authors set out to examine the different ways in which the ethics of algorithms can be normatively questioned. Using their analysis as a starting point, this section of the chapter will examine the normativity of those different types of algorithms and the issues they raise. It will also examine how the use of those algorithms at once undermines our rights and the sovereign control over our autonomy as agents and persons with regard to the control problem of technology. Overall, algorithms are normatively challenging because of the *scale* of analysis, the *complexity* of the decision-making that takes place through the use of AI machine learning techniques, and the *uncertainty* and *opacity* of the work executed by algorithms *autonomously* that results in lack of human control and consequently undermines our own autonomy as users. For autonomy defines and determines personal and collective control and if autonomy is removed by machine learning autonomous algorithms, so is human control (Mittelstadt et al. 2016, 3). That is, our control as users both in our capacity as agents and persons, which takes us back to the problem of control in technology we examined with regard to Stoic and neo-Stoic philosophy in chapters 4 and 7. As Mittelstadt et al., point out (2016, 3–4),

> The algorithm defines decision-making rules to handle new inputs. Critically, the human operator does not need to understand the rationale of decision-making rules produced by the algorithm . . . learning capacities grant algorithms some degree of autonomy. The impact of this autonomy must remain *uncertain*

[emphasis added] to some degree. As a result, tasks performed by machine learning are difficult to predict beforehand ... Uncertainty can thus inhibit the identification and redress of ethical challenges in the design and operation of algorithms.

The lack of human oversight and thus control over machine learning autonomous algorithms, as suggested by the quotation above, is in fact twofold: first, there is a degree of internal loss of control by the Tech companies' human operators; and second, there is external loss of regulatory control and oversight by government agencies, as well by the users themselves whose autonomy as we saw earlier in chapters 4, 6 and 7 is systematically targeted and undermined by the Big Tech companies, such as Facebook and Google, through their persuasive and invasive algorithms that are designed to influence the users' autonomous decision-making choices. The undermining of users' autonomy and control by the Big Tech companies for primarily their self-interest, constitutes therefore a normative violation of the user's rights to freedom, well-being and autonomy, and crucially as persons, as it involves a violation of their absolute right to their dignity, dignity being, as we saw in chapter 7, an absolute human right.

Mittelstadt and his co-authors divide their "map of the ethics of algorithms" into six distinct areas as follows (2016, 4):

(1) **Inconclusive evidence** occurs where the algorithms produce probably, yet inevitably uncertain knowledge, as the statistical methods employed can identify significant correlations but insufficient to indicate a causal connection. This is clearly an epistemic limitation but also an ethical concern as the authors themselves conclude, "the risk of being wrong affects one's epistemic responsibilities"; as a wrongly interpreted correlation may result in harm, as for example, when used in *police profiling*, when a subject is unjustly statistically correlated to a crime they may not have committed. The main epistemic and ethical concerns under the criterion of inconclusive evidence, is that (a) "causality is not established prior to acting upon the evidence produced by the algorithm" and (b) an issue of a crucial methodological concern is that even if, as the authors correctly point out, the correlation or causal knowledge is strong "this knowledge may only concern populations while actions are directed towards individuals." They go on to correctly point out that "actions taken on the basis of inductive correlations [at the population-level] have real impact on human interests independent of their validity" (2016, 5). Is this a case of *ethics by numbers*? This is a crucial question we shall pursue later as it impacts on the very notion of individual autonomy, which is at the heart of most of the issues we will examine in this chapter.

(2) ***Inscrutable evidence***, as we would expect, is identified as major normative issue as it concerns transparency, control, and accountability. For the opacity of algorithms as we have already examined earlier, creates a lack of accountability, as well as lack of control, resulting in lack of oversight, allowing the Big Tech companies to gain control and exercise sovereignty over our individual and collective data. According to Mittelstadt and his fellow authors, the primary components of transparency are accessibility and comprehensibility of information both of which are denied to public view by the Big Tech companies that wish to keep the design and application of their algorithms out of public view. Which, as we have already seen earlier in this book, create an asymmetry of information that favours the commercial and financial interests of Big Tech companies, against the interests of their users and the public good.

The *asymmetry* of information in turn creates an imbalance in knowledge and decision-making power that favours the Big Tech as the curators, possessors and processors of our digital data and consequently to the detriment of our collective normative good, epistemically, ethically and eudaimonically, as individuals and as society. As Mittelstadt and fellow authors correctly observe, "the opacity of machine learning algorithms inhibits oversight" (2016, 6). And without oversight there can be no transparency and consequently no accountability and no control. "Besides being accessible, information must be comprehensible to be transparent" those authors observe (2016, 6). This is an important and crucial point, since given that users, as data subjects, have a legitimate interest to know and understand how the information collected and created about them, and how it influences decisions made by the data-driven practices of the Big Tech companies, they have a legitimate right to transparent and comprehensible knowledge of what algorithms do with their data and how that is done. In short, the asymmetry of information that exists between data subjects and the Big Tech companies, results consequently in a wrongful violation of the data subjects' rights to their autonomy and sovereignty over their digital data. As Mittelstadt et al., correctly point out (2016, 6),

> Meaningful oversight and human intervention in algorithmic decision-making "is impossible when the machine has an information advantage over the operator . . . [or]when the machine cannot be controlled by a human in real-time due to its processing speed and the multitude of operational variables" (Mathias, 2004: 182–183). This is, once again, the black box problem.

The black-box problem to which the authors refer is the problem created by the opacity and incomprehensibility of the operations and rationale of the algorithms, which remain by and large obscure and not readily accessible by the operators as the machine learning algorithms operate as noted earlier, autonomously. Consequently, this inhibits oversight. Both internally, for the operators themselves, and externally, for governments and affiliated regulatory institutions, and society generally. Another crucial problem concerning the inscrutability of algorithms is that they are incomprehensible to most human agents making their legitimacy of decisions difficult to understand and therefore difficult to challenge. Especially, since the asymmetry of information favours the Big Tech companies over the interests of the data subjects in their capacity as users, and society generally. It is an unfair and unjust contest. As Mittelstadt et al., correctly observe (2016, 7).

> Under these conditions, decision-making is poorly transparent . . . and the failure to render the processing logic comprehensible to data subjects, *disrespects their agency* [emphasis added] . . . meaningful consent to data processing is not possible when opacity *precludes risk assessment* [emphasis added].

I have emphasised the sections in the above quotation to highlight their importance to our discussion in this chapter and generally in this book, as both the concepts of *respect* and *risk assessment* play a crucial role in assessing the normativity of ICTs and AI technologies. Respect, as we have already seen in chapter 7 with regard to self-respect, autonomy and dignity, are fundamental rights based on the natural rights to freedom and well-being as instantiated by Alan Gewirth's principle of generic consistency (PGC) and the DOIT-Wisdom model, examined in chapter 5, as well, of course, for they form an integral part of Stoic and neo-Stoic philosophy.

Similarly, risk assessment is crucially important as we shall later see in this chapter with regard to *legislation* and government and societal control over the informational and surveillance practices of the Big Tech companies, which undermine those principles and values. Risk assessment is also crucial in Stoic philosophy, through the Stoic theoretical method of reducing dependence on things outside our human control and focusing only on those things which are within our control, such as our autonomous volition or hegemonikon. However, as we have already seen, our autonomy, dignity and freedom are being systematically undermined by the opaque persuasive technologies of the Big Tech companies. The *risk* to our human autonomy at a time when oversight over the Big Tech companies is presently not only beyond our control but out of control, poses no less than an existential risk whose illegitimacy must be made transparent to bring it within our societal and communal control for the common good of society and the global community. As Shoshanna Zuboff aptly put it, "we have become the *investment futures* in a surveillance culture dominated by the Big Tech companies" (Orlowski, 2019).

Yet another issue concerning the lack of transparency, exacerbated by the unfair asymmetry of information between data subjects and the Big Tech companies, is the matter of *trust*, an issue crucial to a healthy democracy. According to Mittelstadt et al. (2016, 7), "transparency disclosures by data processors and controllers [such as the Big Tech companies] may prove crucial in the future to maintain a trusting relationship with data subjects." In a TED talk (June 2013), Onora O'Neil makes a very pertinent and useful distinction between trust and trustworthiness. For our present purpose, it's what she says about *trustworthiness*, that is most relevant and useful, namely, we should only trust those we know on the basis of good evidence and good judgement, to be *trustworthy*. As she correctly observes,

> Judgment requires us to look at three things. Are they competent? Are they honest? Are they reliable? And if we find that a person is competent in the relevant matters, and reliable and honest, we'll have a pretty good reason to trust them because they'll be trustworthy . . . So, the moral of all this, we need to think much less about trust . . . [and] much more about being trustworthy, and how you give people adequate, useful, and simple evidence that you're trustworthy.

Onora O'Neil intends all three conditions, of competence, honesty, and reliability to be present at once, so as to render one trustworthy. I think from what's been said so far, we would agree that algorithms and their use and control by the Big Tech companies, such as Facebook and Google, although technically competent, cannot be deemed trustworthy, because they lack transparency, accountability, they are not comprehensible, and moreover, there is an informational asymmetry between the Big the Big Tech companies and data subjects that further exacerbates the lack of trustworthiness. Especially, as we now know how their business models, by design and intent, are manipulative and misleading.

That problem as we shall see later in this chapter is further exacerbated and becomes a far more critical existential problem for humanity in the event that AI technology acquires superintelligence, an intelligence that far exceeds that of human beings. That issue will be examined in more detail in chapter 9.

(3) **Misguided evidence** and (4) **Unfair outcomes leading to discrimination**. I have grouped those two issues together as they are closely interrelated. According to Mittelstadt et al. (2016, 7),

> Algorithms inevitably make biased decisions . . . As a result the values of the author [of an algorithm] wittingly or not, are frozen into the code, effectively institutionalising those values . . . social biases can be embedded in system design purposefully by individual designers, seen for instance in manual adjustments to search engine indexes and ranking criteria . . . machine learning algorithms trained from human-tagged data inadvertently learn to reflect biases of the taggers . . . Flaws in the data are inadvertently adopted by the algorithms and hidden in outputs and models produced . . . [and] algorithms also require interpretation [and] explaining the correlation requires additional justification.

The above analysis of the issue of misguided evidence puts paid to the view that algorithms and their use in making statistical correlations that are them applied to individual persons, with sometimes unjust and harmful consequences, are far from being objective and free of biases. At least in the case of human biases, those can be investigated, exposed to public view, and challenged, but not it seems in the case of autonomous algorithms. That should further give us pause and concern about the uncritical use of algorithms without human oversight and control. Mittelstadt et al. observe that "much of the reviewed literature addresses how discrimination results from biased evidence and decision-making" (2016, 8) citing *profiling* by algorithms as one of the main causes, as "they identify correlations and make predictions about behaviour at group-level . . . rather than actual behaviour . . . that can inadvertently create and evidence-base that leads to discrimination." They conclude by pointing out that (2016, 8),

> Bias is a dimension of the decision-making itself, whereas discrimination describes the effects of a decision, in terms of adverse disproportionate impact resulting from algorithmic decision-making . . . [and] the capacity of individuals to investigate the personal relevance of factors used in decision-making is inhibited by opacity and automation.

Once again, what all the issues we have examined so far have in common, is that the use of algorithms, especially by the Big Tech companies, undermine our human autonomy and freedom, as well as our individual control over our behaviour, that the Stoics rightly identified as the one core area in our lives where we could at least exercise control though our rational and virtuous judgements and choices. Now even that internal autonomous control, is accessed and assessed by inscrutable and unaccountable algorithms on the basis of statistical correlations at group-level, which can unfairly impact on us epistemically, ethically and eudaimonically, resulting in a dystopian society not unlike that canvassed in Orwell's novel *1984*. Unlike, Orwell's dystopian vision or nightmare, however, there is still time enough to put a stop to this socially unhealthy and unacceptable situation and challenge the Big Tech companies before it is too late. Our future depends on it.

(5) ***Transformative effects***. A major issue identified under this topic is that personalisation of subjects' data by algorithms poses a threat to their autonomy. This is a crucial issue and one we have examined in terms of Stoic and neo-Stoic philosophy in chapters 3–7 with regard to technology's control over our judgements and decision-making, exercised by Big Tech companies through the use of their persuasive technologies designed to gain control over our autonomy. Once again, we see that control over the autonomy of data-subjects is the overarching critical and crucial issue concerning the impact of the normative effects of algorithms over our well-being as individuals and society generally. As Mittelstadt and co-authors astutely point out (2016, 9),

> Personalization algorithms tread a fine line between supporting and *controlling decisions* [emphasis added] by filtering which information is present to the user based upon in-depth *understanding* of preferences, behaviours, and perhaps vulnerabilities to influence (Bozdag, 2013; Goldman 2006; Newell and Marabelli, 2015; Zarsky, 2016). Classifications and streams of behavioural data are used to match information to the interests and attributes of data subjects. The subject's autonomy in decision-making is *disrespect* [emphasis added] when the desired choice reflects third-party interests above the individual's. (Applin and Fischer, 2015; Stark and Fins, 2013)
>
> The subject can be pushed to make the "institutionally preferred action rather than their own preference" (Johnson, 2013); online consumers, for example, can be nudged to fit market needs by filtering how products are displayed. (Coll, 2013)

An interesting thought is that personalisation algorithms can be taught "to act ethically by striking a balance between coercing and supporting users' decisional autonomy" (2016, 9). However, in theory this might be a practical option to at once reduce coercion and respect the data subject's autonomy, it would not work in practice. For the motivation or the behavioural nudging and coercion of the data subjects' decision-making and choices, is purposely designed by the Big Tech companies into the algorithms on the basis of their business model for the primary objective of psychologically influencing the data subjects' autonomy.

As Mittelstadt et al. correctly observe,

> Filtering algorithms [through personalisation] that create "echo chambers" devoid of contradictory information may impede decisional autonomy

(Newell and Marabelli, 2015). Algorithms may be unable to replicate the "spontaneous discovery of new things, ideas and options" which appear as anomalies against the subject's profiled interests. (Raymond, 2014)

Even though as Mittelstadt et al. suggest, the personalisation algorithms might "both enhance and undermine the agency of data subjects, and thus their autonomy," the crucial issue is that the subjects' data autonomy is not completely within their control and given the informational asymmetry that presently exits between the Big Tech companies and data subjects, it is not clear how much control data subjects have over their autonomy. Certainly, according to the Stoics not enough and not sufficient in ensuring one's control over their autonomy and prohairesis and thus one's well-being or eudaimonia, which then becomes hostage to someone else's will and direction.

Another significant and major issue to be examined now in this section is that of *privacy*, for as Mittelstadt et al. correctly observe "algorithms are also driving a transformation of notions of privacy . . . A right to identity derived from informational privacy interests suggests that opaque or secretive profiling is problematic (2016, 10). In an important and timely new book *Privacy is Power*, Carissa Véliz (2020) correctly observes that,

> Surveillance threatens freedom, equality, democracy, autonomy, creativity, and Intimacy. We have been lied to time and again, and our data is being stolen to be used against us. No more. Having too little privacy is at odds with having well-functioning societies. Surveillance capitalism needs to go. It will take some time and effort, but we can and will reclaim privacy (2020, 5) . . . Aristotle argued that part of being virtuous is all about having emotions that are appropriate to the circumstances. When your right to privacy is violated, it is appropriate to feel moral indignation. It is not appropriate to feel indifference or resignation . . . Privacy is a right for good reasons, defend it. (2020, 201)

Véliz is of course right in her insightful and polemic analysis. Clearly, for the Stoics, as for Aristotle, at least with regard to having good reasons, the control of one's informational privacy is a central part of what it is to be a fully autonomous agent capable of exercising *internal control* over one's judgements and choices made on the basis of one's own volition or prohairesis as we saw in chapter 7. The idea that third parties can access one's informational privacy especially without one's informed consent, and not uncommonly without their knowledge, would be considered a fundamental violation of the central principle of Stoic and neo-Stoic philosophy. For since external circumstances are not in our control, our judgements, and choices, our volition as free autonomous agents and self-respecting persons, as the only things within our human control, should always be within our control as they have a direct impact on our well-being and eudaimonia. To allow third parties to share control over our informational privacy would amount to surrendering control over ourselves, and hence our freedom, dignity, and integrity, consequently allowing others to determine what our well-being ought to be according to their interests, not ours. Moreover, as we saw in chapter 7 with regard to the neo-Stoic philosophy that I am attributing to Alan Gewirth through my Stoic-reconstruction of his concept of self-fulfillment (see chapter 3), a violation of one's informational privacy would constitute a violation of one's rights to freedom and well-being as both agent and person.

As Mittelstadt et al., observe,

> Identity is increasingly influenced by knowledge produced through analytics that makes sense of growing streams of behavioural data. The 'identifiable individual' is not necessarily a part of these processes . . . Profiling seeks to assemble individuals into meaningful groups, for which identity is irrelevant (Floridi, 2012; Hildebrandt, 201; Leese, 2014). Van Wel and Royakkerss (2004:133) argue that external identity constructions by algorithms is a type of *de-individualisation* [emphasis added] or a "tendency of judging and treating people on the basis of group characteristics instead of on their own individual characteristics and merit" . . . The individual's informational identity (Floridi, 2011) is breached by meaning generated by algorithms that link the subject to others within a dataset (Vries, 2010) . . .
> A lack of recourse mechanisms for data subjects to question the validity of algorithmic decisions further exacerbates the challenges of *controlling identity and data about oneself.* (Schermer, 2011) [emphasis added]

Needless to say, the above astute observations and comments made by Mittelstadt and co-authors raise some very important and crucial normative issues concerning autonomy, self-sovereignty, identity, privacy, freedom, and especially control, which, as previously emphasised in this chapter and chapter 7, is the overarching normative issue with regard to the external control exercised by the Big Tech companies over us; as well as the internal control we as individuals ought to have over our own autonomy, freedom, dignity and integrity, as both agents and importantly as persons owed unconditional respect and an absolute right to our dignity.

As clearly indicated in the above quoted passage, our individual identity is under threat by the digitalisation and datafication of our identity through opaque, incomprehensible and unaccountable algorithms that construct our digital identities without our knowledge nor consent, based on statistical data gathered at group level that has very little, if anything, in common with our natural individual identities. And yet our group-level identities are the ones that are used as the basis for making decisions that affect our natural identities that can result through impersonal profiling to unjust outcomes for our individual natural selves.

In a paradoxical turn, our digital identity is and can be set in normative opposition and seemingly in conflict with our natural identity, when through profiling someone may be deemed guilty of a crime presumed to have committed or at least capable of having committed, or about to commit, on the basis of the profile of their *digital group-level identity*, which their natural identity is unaware of having committed.

Yet another insidious and potentially crucial existential problem not just for data subjects but humanity generally, is the split of our identity into two separate identities: a *group digital identity* controlled externally by algorithms, and our *natural individual identity* controlled internally by us as individuals, but whose autonomy, however, is also precariously under the intended systemic control by the persuasive technologies of the Big Tech companies, such as Facebook and Google.

For those of us acquainted with *Star Trek*, this issue of opposing and conflicting identities, our *digital group identity* (DGI) vs. our *natural individual identity* (NII) should remind us of the collective digital identity of the *Borg*: an alien species that conquers other planets and 'assimilates' its citizens within its digital collective hive, without the consent nor willingness of their victims, that are forcefully assimilated into the collective digital hive, consequently losing their own personal and individual identities. When the Borg meet resistance by their potential victims, their warning is *Resistance is Futile*. In the Star Trek series *Voyager*, *Seven-of-Nine*, a human young woman assimilated by the Borg when still a child is rescued by the human crew of Voyager and restored to her original individual identity but not before much difficult changes and adaptations that *Seven-of-Nine* has to face in learning how to become an individual human person again, with one's own thoughts, judgements, choices, and volition. One could say, at least metaphorically, that "Seven- of-Nine's" case is not only one of *identity theft* but also one of *identity-amnesia* one suffers when one is assimilated into a collective identity hive, through the digitalisation and dehumanisation of their natural personal identity, that denies them their own natural individual identity.

This raises a paradox: That the autonomy of our NII is systematically and methodically usurped by the Big Tech companies through their persuasive technologies, through their autonomous algorithms, that are used through profiling to create our DGI, which is then applied to further undermine our digital individual identity and consequently our autonomy that can result in unfair and unjust outcomes (see sections 3 and 4 above) resulting from statistical group-level profiling. For the Big Tech companies, it is a case of having their cake (their algorithms) and eating it too (our data). As Shoshana Zuboff astutely points out (2019), in the surveillance capitalism era, *we are the product*. A very valuable data product, a commodity that the Big Tech companies sell to third parties, to make themselves extremely wealthy, very powerful and very unaccountable.

(6) ***Traceability***. Under this final section, Mittelstadt et al. examine the traceability of algorithms and the moral responsibility attributable to them (Mittelstadt et al. 2016, 10–11). The overarching normative issue examined in this final section, just as the other issues examined under the previous sections above, concern the *control problem of technology* and specifically control over our individual autonomy as data subjects, in our dual capacity, as both agents and persons. Significantly, in this section the main issues of crucial concern raised, are: who is normatively responsible for the decision-making and choices of algorithms, given their incomprehensibility and lack of transparency and oversight; should autonomous or semi-autonomous machine learning algorithm be held wholly or partly normatively responsible for the decisions and choices they make, and who should provide that oversight of accountability? Should machine learning algorithms be designed and learn to behave ethically? What should the normative relationship be between our *group-level digital identity (DGI)* and our *natural individual identity* (NII)? If both identities are controlled by the algorithms of the Big Tech companies, at least to some extent, what are the normative consequences for our autonomy, for both our digital and natural autonomy, and how and who should provide oversight that our individual rights as agents and persons are protected against the manipulation and exploitation by the Big Tech companies? It was deemed necessary to highlight the various issues raised by Mittelstadt and his co-authors. In summary, these are: ethical responsibility for algorithms – who is responsible; Human oversight and accountability – how and who should implement it; The problem of the designer's control and the machine learning algorithm's autonomous behaviour that creates an *accountability gap*; Should autonomous algorithms be held ethically responsible and accountable; Should the moral responsibility concerning algorithms be a distributed-shared responsibility between human operators and algorithms; Should moral reasoning and behaviour be designed into algorithms and AI artificial agents, and how?

As indicated above throughout this chapter, the overarching and core theoretical and practical normative problem of the aforementioned summarised issues, is the crucial normative issue of *control*. Namely, who should be in control of autonomous algorithms and AI artificial agents – *Us* individually and collectively as data subjects and citizens and society generally, or the **Big Tech** companies? As we examined earlier in this chapter and chapter 7 in relation to Stoic and neo-Stoic philosophy, the problem of control concerns no less than our autonomy as human agents and persons and consequently our well-being that is dependent on having at least self-control over our own autonomous judgements, decisions, and choices. That control, as we examined in detail in chapter 7 and presently in this chapter, shows that given the informational asymmetry that exits between *Us* and **Big Tech** companies, as well as the systemic targeting and undermining of our autonomy by the persuasive technologies and algorithms of those companies, our individual and collective control over our *human autonomy* is in jeopardy. Added, to the aforementioned issues, the following section will address and examine the control problem concerning autonomy in relation to what I refer to in this chapter as the *identity-paradox*; an additional, though implied, issue and problem to those referred above.

THE IDENTITY-PARADOX

I initially explored the related and crucial issue of the *identity-paradox* or *conundrum* in subsection Transformative effects above, with regard to the *transformative* effects of algorithms. Namely, that our individual identity is under threat by the digitalisation and datafication of our identity through opaque, incomprehensible, and unaccountable algorithms that construct for us digital identities without our knowledge nor consent, based on statistical data gathered at group level that has very little, if anything, in common with our natural individual identities. Yet, our group-level identities are the ones that are used as the basis for making decisions that affect our natural identities that can result through impersonal profiling to unjust outcomes for our individual natural selves (see subsection Unfair outcomes leading to discrimination above). In a paradoxical turn, our digital identity is and can be set in normative opposition and seemingly in conflict with our natural identity, as when through profiling someone may be deemed guilty of a crime presumed to have committed, or at least capable of having committed, or about to commit, on the basis of the profile of their digital group-level identity, which their natural identity is unaware of. Moreover, yet another insidious and potentially crucial existential problem, not just for data subjects but humanity generally is the following: the split of our identity into two separate identities, that is, a *group digital identity* controlled externally by algorithms, and our *natural individual identity* normally controlled internally by us as autonomous individuals, is also, however, precariously now under the systemic manipulative control of the persuasive technologies of Big Tech companies, such as Facebook and Google. Thus, both our digital and natural identities are increasingly placed under the outside self-interested control of Big Tech companies and other third parties often without our knowledge and consent. From a Stoic and neo-Stoic perspective, this is normatively unacceptable as it violates our fundamental autonomy as agents and persons, to which, as we saw in chapter 7, we have prima-facie rights as agents and an absolute right as persons. By applying the dual-aspect of personal identity and agency, supported theoretically by both Kant's and Gewirth's rational arguments as we saw in chapter 7, as well as the principles of Stoic and neo-Stoic philosophy, we can now demonstrate how the problem of the *identity-paradox* created by the algorithms of Big Tech companies at once undermines the data subjects' natural individual autonomy, as well as their group digital-autonomy, and therefore, breaches and violates the data subjects' dignity in a twofold manner, that is, as both digital-agents and individual persons.

The normative harm to data subjects and generally to society, as the collective unity of individual persons qua citizens, is therefore double. This brings us now to the point of considering what must be done to remove or at least mitigate such unacceptable harm to our dignity as agents and persons. According to Mittelstadt et al. (2016, 11–12), "automation of decision-making creates problems of ethical consistency between humans and algorithms."
The Stoics would agree of course as consistency of virtuous conduct was for them a necessary principle for ensuring a person's *integrity*, an essential aspect of one's normative autonomy and *hegemonikon* or volition, for making the right decisions and choices, thus contributing to a person's well-being. Stoic and neo-Stoic philosophy with its emphasis on rationality, virtues and logical arguments and principles, is well placed to make an important and valuable contribution to that ongoing debate, as we examined in chapters 3 and 4, with the Gewirthian neo-Stoic philosophy's emphasis on natural rights to freedom and well-being formulated around the dual concepts of agenthood and personhood, centred on the value of dignity, which was examined in chapter 7 under the section, *The Dignity-Conferring Value of Rights*.

The significance of this is that the neo-Stoic model of philosophy presented in this book, on the basis of Gewirth's rationalist ethics of rights (see chapters 3 and 4), allows us to define and evaluate the normative impact the incursions and violations of Big Tech companies have on our dual autonomy, in terms of both *virtues and natural rights*. This helps to provide a more nuanced evaluation of the ethics of algorithms that take into account different types of normative evaluative theories, such as rationalist ethics, virtue ethics and eudaimonic ethics, as those of the Stoic and neo-Stoic theories advocated in this book, which combine in one theory, all three of those different approaches to ethics, that is, natural right ethics, virtue ethics, and importantly eudaimonic ethics, which as we saw also feature in the DOIT-Wisdom normative model examined in chapter, 5, of this book. Moreover, the difficulty highlighted by Mittelstadt and his co-authors, namely, that "consensus view does not yet exist for how to practically relocate the social and ethical duties displaced by automation" (2016, 11–12) seems slightly exaggerated and can be overcome since on a closer examination most ethical theories, such as utilitarian, contractarian, communitarian and ethics of care theories, as well as rationalist theories such as those of Gewirth's and Kant's, do converge on key ethical principles, such as equal rights, dignity, mitigation of harm, promotion of the collective common good, duty of care, equality and respect and dignity.

Those theories would therefore assess the invasive and manipulative practices of the Big Tech companies that systemically undermine our autonomy and the common good of society as normatively problematic, dis-demonic and harmful. For as we have seen so far, in this book, the Big Tech companies do systemically undermine and violate our epistemic, ethical and eudaimonic rights and our common good and well-being as individuals and society.

SHARED RESPONSIBILITY, COMPLIANCE AND LEGISLATION

Two final issues raised by Mittelstadt and his co-authors, concern (a) traceability of *shared and distributed ethical responsibility* between human operators and autonomous algorithms, and consequently (b) accountability and ethical compliance. According to those authors (2016, 12–13),

> Substantial *trust* [emphasis added] is already placed in algorithms, in some cases affecting a *de-responsibilisation* of human actors, or a tendency to "hide behind the computer" and assume *automated* processes are correct by default (Zarsky, 2016: 121). Delegating decision-making to algorithms can shift responsibility away from human decision-makers . . . additional research is needed to understand the prevalence of these effects in algorithm driven-making systems, and to discern how to minimise the inadvertent justification of harmful acts (Davis et al., 2013) . . . and further work is also required to specify requirements for resilience to malfunctioning as an ethical ideal in algorithmic design.

The two main crucial problems of the trust placed in autonomous algorithms is that their opaqueness and inscrutability, as well as the information asymmetry that exists between data subjects and the Big Tech companies that exercise control over those algorithms, render them as we saw earlier, *untrustworthy,* and therefore our trust as data subjects and users and society generally, should not be uncritically placed in such autonomous algorithms, over which we have little control as human operators and no control as human users. Moreover, we should not willingly surrender our own individual and collective normative responsibility over the harmful effects of algorithms on human agents, as to do so, as Mittelstadt et al., correctly point out, leads to *de-responsibilisation* of human actors and by extension, I wish to add, can lead to the de-facto *dehumanisation* of our humanity us both agents and persons, and society generally. For it removes control over our autonomy, and our own responsibility in protecting that autonomy as a *fundamental right*, both as individuals and collectively as societies. Autonomy as the Stoics and neo-Stoics correctly argue (see chapter 7) is a fundamental characteristic of what it means to be human, as agents and persons. Moreover, delegating our own autonomy irresponsibly or involuntarily is not in our best interest, for as the Stoics correctly point out, autonomy is the only thing we have within our control. Therefore, we should retain overall control over our own autonomy, as its loss or reduction impacts directly on our well-being, and hence would at least constitute a dereliction of care of duty by us. Loss of control over our autonomy as individuals and societies, due to the systemic invasive and manipulative practices of the Big Tech companies through their unaccountable opaque algorithms would also constitute a violation of our fundamental absolute rights to our dignity as persons, as we saw earlier in chapter 7.

With regard to transparency and ethical compliance, Mittelstadt and his co-authors correctly point out (2016, 12–13), "how to operationalise transparency remains an open question, particularly for machine learning." They go on to say,

> That merely rendering the code, of an algorithm transparent, is insufficient to ensure ethical behaviour . . . One possible path to explainability is *algorithmic auditing* [emphasis added] carried out by data processors (Zarsky, 2016), external regulators (Pasquale, 2015) and (Tutt, 2016); or empirical researchers (Kitchin, 2016; Nayland, 2016), using ex post audit studies (Adler et al., 2016) . . . or reporting mechanisms designed into the algorithm itself (Vellido et al., 2012). For all types of algorithms auditing is a necessary precondition to *verify* correct functioning . . . it will require cooperation between researchers, developers and policy-makers.

Auditing, both internally by the companies that use algorithms, such as the Big Tech companies, and externally by government and independent third-party regulators through legislation, is necessary as Mittelstadt and his co-authors correctly indicate above, to ensure *ethical compliance*. Regarding the necessity for algorithmic transparency and ethical compliance, the aforementioned cited comments by Mittelstadt and co-authors are in keeping with what I argue in my book (Spence, E., 2021), *Media Corruption in the Age of*

Information. In summary, the following are some of the main issues raised (Spence, E., 2021, Chapter 7: 15–27).

In an op-ed article by Mark Zuckerberg, "Big Tech Needs More Regulation" published in the *Financial Times* (February 16, 2020), the Facebook CEO argues for more regulation along similar lines as his previous article published on *Facebook* on March 30, 2019. This is a welcome initiative by Facebook and in keeping with the suggestion canvassed above in this chapter and chapter 6, that the protection and safeguarding of the normative integrity of information and its communication on the infosphere, especially its dissemination through the Big Tech platforms, such as those of Facebook and Google, should be a shared responsibility: a collective normative responsibility, epistemic, ethical and eudaimonic, shared collectively by all relevant key stakeholders, including tech companies, government, market actors, the media and citizens as users and all communicators of information generally. As Mark Zuckerberg correctly states,

> People shouldn't have to rely on individual companies addressing these issues [regulation issues] by themselves. We should have a broader debate about what we want as a *society* [emphasis added] and how regulation can help . . . It's time to . . . define clear responsibilities for people, companies and governments going forward. (Zuckerberg, 2019)

This is a belief also expressed persuasively by the authors of *The Platform Society* (Van Dijck, Poell and De Waal 2018, 139), who claim that,

> Creating public value for the common good should ideally be the shared responsibility of market, state, and civil society actors. The Big Five [Facebook, Google, Amazon, Apple, and Microsoft], as major shapers of the ecosystem, can rewrite the rules for democratic societies; but how can they act responsibly in a new global world order . . . Public institutions have a special function as *guardians* [emphasis added] of social trust and a democratic commons.

The three authors go on to correctly suggest that, "platformization requires an integral approach rather than just a sectoral one when it comes to defending consumer's and citizens' rights" (2018, 139).

The comments of Mark Zuckerberg and those of the authors of *The Platform Society* cited above, provide further support for a shared and integral global approach to the normative regulation of the Internet as argued for in this chapter. It should include the adoption and inculcation of public ethical principles and values and in addition, government regulation to ensure compliance with those principles and values. The DOIT-Wisdom theory, we examined in chapter 5, was designed specifically for the purpose of providing an information-communication normative model that could apply globally. For the normative principles applicable to the Internet, being a global network, requires a normative model that has global applicability.

Another important area concerning regulation that Mark Zuckerberg highlights in his *Financial Times* co-ed article (2020), is *oversight* and *accountability* that goes beyond mere self-regulation and clearly and unambiguously introduces the need for *external* government regulation. As he states in that article,

> people need to feel that global technology platforms answer to someone, so regulation should hold companies accountable when they make mistakes.

He goes on to say

> that if we don't create standards that people feel are *legitimate* [emphasis added], they won't trust institutions or technology. (Mark Zuckerberg, 2020).

This statement by Zuckerberg is significant for at least two important reasons:

1. The statement suggests not only retrospective regulation through substantial fines imposed on tech companies such as those imposed on Facebook and Google for their past "mistakes," but crucially, *prospective regulation* through regular and ongoing "oversight." The term "oversight" suggests proactive ongoing regulation, just as, for example, in the case of annual independent auditing of financial statements and tax returns.
2. Zuckerberg, in fact, suggests this interpretation of proactive regulation through oversight in his 2019 article cited earlier, "Four Ideas to Regulate the Internet," in which he highlights the "need for a more active role for government and regulators." Later in that article he refers to the *transparency reports* Facebook publishes on its effectiveness of removing harmful content on its platform, which he believes every Internet service should provide, and which he considers "just as important as financial reporting."

As a professional practising accountant and auditor for many years both in the UK (London) and Australia (Sydney) in my previous working life before becoming an academic philosopher, Mark Zuckerberg's statement above "just as important as financial reporting" caught my immediate attention and interest. For as an independent accountant specialising in auditing, I conducted external audits of many private and public corporate companies, as well as external audits of government departments. This essentially involved examining not only the veracity of the annual financial statements prepared by those companies but also the internal underlying systems of those companies on which the reporting of their financial statements was based. For if, for example, the stock accounting system of a company was faulty or inaccurate the reported stock value of that company in their reported financial statements, which determined the annual profit or loss of the company, would also be inaccurate and therefore unreliable and untrustworthy.

Analogously with regard to the internal systems of tech companies, such as Facebook and Google, for example, that collect, commodify, select and monetise user's information that generate their reported profits, an independent external audit would seek to verify that process to ensure it was conducted according to accepted normative public principles, values and standards such as those cited above, namely, *privacy, transparency, accuracy, trustworthiness, responsibility* and *accountability,* which respects their users' normative interests and those interests are fairly balanced with, and not overridden by, the financial interests of the tech companies. Such an independent external *normative audit* of tech companies would provide a reasonably effective oversight in the form of external regulation, which would parallel the existing financial audits of most private and public corporate companies at present. Referring to the Big Tech companies' opaque algorithms that for the most part remain non-transparent and not subject to public accountability, Frank Pasquale's comment in his book *The Black Box Society* (2015) seems to lend support to my proposal for external platform audits of Big Tech companies such as Facebook and Google. He says that, "far more of their algorithms should be open for inspection – if not by the public at large, at least by some trusted auditor" (Pasquale, 2015, 141).

In his article "Four Ideas to Regulate the Internet," Mark Zuckerberg (2019) also hints at such a possibility by stating that,

> One idea is for third-party bodies to set standards governing the distribution of harmful content and measure companies against those standards. Regulation could set baselines for what's prohibited and require companies to build systems for keeping harmful content to a bare minimum.

Although Zuckerberg's statement refers specifically to harmful content, there is no reason, in principle at least, why that could not also be extended to include issues of privacy and data protection, as well as the protection of elections through the monitoring of political ads, which he also refers to in that article. The general ideas that Mark Zuckerberg expresses in that article is also in keeping with the overarching *normative* statement he makes in his 2020 *Financial Times* article, that "tech companies should serve society."

This is an important and significant statement by the CEO of Facebook as it emphasises and acknowledges, by implication at least, the normative priority of the common good of society over the narrower financial interests of tech companies, which as illustrated by the Cambridge Analytica case in chapter 6, don't always serve the common good of society.

As we saw in chapter 2 of this book, technology is not good in itself but only good if it serves the collective common good of societies and humanity at large, which lends theoretical and normative support to Mark Zuckerberg's statement that technology should serve society.

The idea of an independent audit of tech companies to assess compliance with common standards comprising public normative principles and values, as proposed above, faces, however, the challenge of adequate expertise and professional competence by the external regulators (to whom I will refer to here as *platform auditors*) in carrying out such a demanding role. Although not referring to independent external auditors as I have done above, the authors of the *Platform Society* outline some of the problems that enforcement agencies in general would face in carrying out such a regulatory role. They claim, not unreasonably (2018, 158) that,

> Enforcement agencies that are assigned to execute regulation are *insufficiently equipped to* administer even the most basic form of accountability. Regulatory fixes require detailed insights into how technology and business models work, how intricate platform mechanisms are deployed in relation to user practices, and how they impact social activities . . . Government regulators need specialized digital teams equipped with combined technical and legal knowledge to probe these complex mechanisms.

Although the challenge as described above for introducing external regulation of tech companies through external compliance audit is difficult, it is not insurmountable. Large traditional financial audits of national and international companies conducted by independent external auditors, working for professional accountancy firms, such as those of the Institutes of Chartered Accountants in the UK and Australia, for example, are not without complexity and also do require specialist adequate knowledge of the systems of the companies audited. My own experience as audit manager, working with a team of auditors, required us to liaise closely with the managers and staff of the various sections within each company we audited, so as to first determine what the systems of each section of the company were and how those systems functioned. Once we acquired an overall understanding of a company's systems, which included its computer systems, we would then devise audit checks and controls to test those systems for their functionality, veracity and adequacy. Admittedly, the algorithmic systems of tech companies such as Facebook and Google might be much more complex but auditors in the future conducting regulatory audits of tech companies would be expected and specifically trained to have a basic knowledge of how the machine learning computer systems of tech companies function so as to enable them to at least test them. This might require the platform audit teams in the future to include auditors that have adequate knowledge of computers and their AI systems, including their algorithms.

If compliance with public normative principles and values as those cited above by the Big Tech companies is essential for the common good of society and its underlying democratic system, as well as its political and civil associated institutions, then some government external regulation whether conducted through independent external auditors as suggested above, or some other effective way, is not only desirable but necessary. Mark Zuckerberg's acknowledgement of the necessity of external government regulation is therefore a significant first step in motivating such a system of external government regulation. As argued above, good governance of the Internet requires a joint collective effort by all stakeholders in society, including tech companies, market actors, governments, and citizens, and all the users of the Internet in fact. Historically, governments through the political power assigned to them on the basis of the social contract for the common good of society, have the authority and jurisdiction to introduce regulatory systems to protect the rights of citizens and consumers.

The Information Commissioner's Office (ICO) in the UK, the Federal Trade Commission (FTC) in the USA, as well as the European Union (EU) through the introduction of the *General Data Protection Regulation* (GDPR) have all imposed hefty fines on both Facebook and Google for their misconducts. In October 2019, the *Australian Competition and Consumer Commission* (ACCC) took Google to court for misleading customers about its use of personal data. According to the ACCC Chairman Rod Sims, the ACCC was,

> Taking court action against Google because we allege that as a result of these on-screen representations, Google has collected, kept, and used highly sensitive and valuable personal information about consumers' location without them making an informed choice. (Hunter, 2019)

The court action emerged from the ACCC's released digital platforms inquiry in July 2019, proposing drastic changes to regulation of the Big Tech companies, including stronger protection for consumer data. This is how the Treasurer of the Australian Government Josh Frydenberg described the proposed regulation of the ACCC,

> Make no mistake, these companies [Facebook and Google] are among the most powerful and valuable in the world and they need to be held to account and their activities need to be more transparent. (Hunter, 2019)

Such government legislation at least across the three continents, the United States, the EU and Australia is a significant step forward in the regulation of the Big Tech companies, especially Facebook and Google, which between them, through the various activities of their platforms, exert enormous monopolistic global power over all sectors of society, social, political and financial. Bringing the Big Tech companies under institutional oversight for accountability through government regulation is a positive measure for ensuring the public common good of society is protected and promoted, as acknowledged by Mark Zuckerberg in his two cited articles above.

In addition to government regulation, civil society actors as citizens have a collective shared responsibility to protect and promote the normative integrity of the infosphere just as they do or ought to do, of the biosphere. This as we saw earlier in chapter 2 as well as in chapter 5, is also normatively required by the DOIT-Wisdom theory which demonstrates that the dissemination and communication of digital information in the infosphere commits all of us globally, as disseminators and communicators of information, to epistemic and ethical principles and values. As the authors of the *Platform Society* correctly observe, the platformisation of society has introduced "a new social contract, based on trust and cooperation which sees citizens and consumers not as obstacles or resources to be exploited, but as partners" working together for the common social good of society (Van Dijck et al., 2018, 151). According to those authors,

> Digital platforms can empower individual citizens to unite and to rally behind public advocacy efforts in order to influence democratic processes . . . individuals can muscle up collective power and become a force to be reckoned with because user metrics are one of the biggest currencies of the platform economy.

The empowerment of citizens to collectively protect the privacy and integrity of their information and its use by Tech companies such as Facebook and Google, is not only desirable, but moreover, as we examined in chapter 6 of this book, a *normative responsibility* to be shared collectively by the users of platforms such as those of Facebook and Google, to safeguard the integrity of their information and prevent its misuse. For as we saw, in the Cambridge Analytica scandal, users' information can be misappropriated by the actions of Tech companies such as Facebook and Google, which in some cases, have been shown to be conductive and result in systemic media corruption. Such actions not only undermine the users' informational rights as individuals but also undermine our collective rights as citizens, which also, in turn, undermine democracy itself and its associated and supportive institutions.

Tim Berners Lee, one of the early pioneers of the Internet and creator of the *Open Society Foundation*, has supported a democratic, fair, and secure Internet where citizens can exercise democratic control over information and data flows. When presenting the goals of the *Open Society Foundation* in 2017, he raised three key concerns regarding the future of the World Wide Web (Web) as a tool in the service of humanity. Those were: the *loss of control* over our personal data; how easily misinformation can spread on the Web; and the opaque ways political advertising campaigns are directly targeted at users, as illustrated by the Cambridge Analytica case (Van Dijck et al., 2018, 153). Tim Berners Lee's three key concerns for the Internet provide further motivating reasons for civil society actors, together with public and government institutions, to collectively work together towards a more effective regulation of the Big Tech companies for the common social good of society. To that end, the participation of the tech companies themselves, as suggested by the CEO of Facebook Mark Zuckerberg, is also vital. As Van Dijck et al. (2018, 162) correctly argue,

> A digital world in which large corporations have both an overwhelming market presence and the leverage to influence political actors gives rise to highly unbalanced *polities*. For democracies to work in the age of platformization, they need the concerted efforts of all actors – market, state, and civil society – to build a sustainable and *trustworthy* [emphasis added] global platform ecosystem, a system that comes equipped with distributed responsibilities as well as with checks and balances.

DIGITAL RIGHTS AND THE RIGHTS TO IDENTITY

As we saw above, Gewirth's PGC argument for universal natural rights to freedom and well-being, on the basis of which in chapter 3, I propose a neo-Stoic philosophy that is centred around Gewirth's theory of self-fulfillment and self-respect, establishes that agents *qua* persons have not only prima facie rights as agents, but moreover, have absolute rights to their dignity or self-respect. The dignity-conferring value of absolute rights is both crucial and significant for it demonstrates that not only Big Tech companies, such as Facebook and Google, undermine peoples' personal autonomy but moreover, violate the absolute right people have to their dignity as persons, as well as the rights they have to their freedom and well-being on which their dignity necessarily depends. As we saw in chapter 7, it is because of this *intrinsic* relation between, on the one hand, freedom, and well-being as the objects of the generic rights, and on the other, freedom and well-being as the essential and fundamental constituents of the personal property of dignity, that the generic rights of agents, has a *dignity-conferring value*. That is, they have a dignity-conferring value precisely because the objects of those rights, namely, freedom and well-being, are simultaneously the essential and fundamental constituents of the property of dignity. If freedom and well-being were merely valued as the necessary instrumental *means* for action, and not also valued as the essential and fundamental constituents of an agent's dignity, then the generic rights of an agent would not have the dignity-conferring value, to which Gewirth refers.

Insofar as dignity is a fundamental and necessary property of a person's identity, for as we examined in chapter 7, people would at times choose to end their lives rather than compromise or surrender their personal dignity as the only way of preserving it, then dignity is also an essential valued characteristic of one's individual identity. Notice that although a person may choose to die for their dignity they do so precisely because they wish to preserve their dignity as it is an essential and fundamental characteristic of *their identity*, which they wish to preserve and be remembered for, even after death. That is, as a person of dignity and integrity, worthy of respect. It is for such rational reasons that the Stoic Roman senator Cato chose to end his life, rather than surrender to Julius Caesar and ask for pardon, a pardon Caesar was willing to grant, after the defeat of the Republican army under Cato's command. For Cato knew that surrendering to Julius Caesar and accepting Caesar's pardon would be an unacceptable compromise to his own Stoic principles and those of the Roman Republic, for which he was prepared to die. As we saw above (see the the Identity-Paradox section), according to Gewirth (1985, 743),

> The ultimate purpose of the [generic] rights is to secure for each person a certain fundamental moral status: that of having rational autonomy and dignity in the sense of being a self-controlling, self-developing agent who can relate to other persons on a basis of mutual respect and cooperation, in contrast to being a dependent, passive recipient of the agency of others.

And similarly in the same spirit, Kant claims (1981, 43),

> That merely the dignity of humanity as rational nature . . . should constitute the sublimity of maxims and the worthiness of every rational subject to be *a legislative member*[emphasis added] in the kingdom of ends.

By applying the dual-aspect of personal identity and agency, supported theoretically by both Kant's and Gewirth's aforementioned rational arguments, as well as the principles of neo-Stoic philosophy that also incorporates

Gewirth's justification for generic rights to freedom and well-being, we can now conclude that data subjects, as both agents and person, have prima facie generic rights to freedom and well-being as agents and absolute rights as persons. For as we saw in chapter 5, the DOIT-Wisdom model demonstrates the dual-normative structure of *informational action,* to which all informational agents, both analogue and digital, are committed by universal necessity. Information generally can thus be epistemically, ethically and eudaimonically evaluated *internally* by reference to its inherent normative structure. That structure commits all disseminators of information, to ethical, epistemic and eudaimonic universal rights for all agents, both digital and analogue. Therefore, the Big Tech companies are by virtue of the DOIT-Wisdom model also committed to respecting both the analogue and digital rights of data subjects in their dual capacity as both agents and persons.

Regarding the data subjects' rights concerning algorithms, the EU GDPR stipulates a number of responsibilities of data controllers and rights of data subjects. According to Mittelstadt and co-authors (2016, 13–14),

> Concerning the former, when undertaking profiling controllers are required to evaluate the potential consequences of the data-processing activities via a data protection impact assessment (Art. 35(3)(a)). In addition to assessing privacy hazards, data controllers also have to communicate these risks to the persons concerned. According to Art.13(2)(f) and 14(2)(g) data controllers are obligated to inform the data subjects about existing profiling methods, its *significance* and its *envisaged consequences.*

Those authors, however, correctly anticipate and recognise (2016, 14) that

> Additional work is required to provide normative guidelines and practical mechanism for putting the new rights and responsibilities into practice . . . and an appropriate balance between data subjects' rights to be informed about the *logic* and *consequences* of profiling, and the burden imposed on data controllers.

The rights of data subjects in accordance with *EU GDPR* regulations is both significant and timely and is in keeping and parallel to our aforementioned discussion of the rational and moral justification of the universal rights of data subjects, in their capacity as both agents and persons, with dual rights to their dignity and digital identity, on the basis of Gewirth's argument for the PGC, the DOIT-Wisdom model, and the Gewirthian neo-Stoic theory we examined in chapters 3–5.

Of particular importance is the ability to demonstrate, as we have done, in this chapter and chapter 7, that a Gewirthian neo-Stoic theory incorporates and is able to apply a theory of universal rights rendering that neo-Stoic theory more compatible and relevant to contemporary times, as a fully-fledged notion of rights, is not present nor does it methodologically form part of traditional Stoic philosophy. This is a unique feature of the Gewirthian neo-Stoic theory canvassed and supported in this book as it combines both virtues and rights that renders that theory consistent with both virtue ethics and rights-centred rationalist ethics. With regard to rights to identity, Mittelstadt and co-authors also agree in keeping with our analysis above, that "data subjects can be considered to have a right to identity" (2016, 16). They point out that,

> Such a right can take many forms, but the existence of *some* right to identity is difficult to dispute. Floridi (2011) conceives of personal identity as constituted by information. Taken as such, any right to informational privacy translates to a right to identity by default, understood as the right to manage information about the self that constitutes one's identity . . . and can be connected to the right to personality derived from the European Convention of Human Rights.

Interestingly, notice that Floridi's conception of identity as one constituted of information is normatively aligned, as least theoretically, with the DOIT-Wisdom model that derives the normativity of information internally on the basis of what information is by virtue of its inherent constitutive characteristics as communication. Thus, digital identity understood as constituted by information, would also analogously possess its normative characteristics; that is, epistemic, ethical and eudaimonic. Hence, the violation of one's right to one's digital identity, which as we saw above, we related to one's dignity, also constitutes a violation of its constituent parts, namely, the epistemic, ethical and eudaimonic. Therefore, as we saw earlier in this chapter and chapter 7, the normative impact of deception and manipulation on our digital identity by Big Tech companies, through their persuasive opaque technologies, adversely affects us as data subjects and society generally, in a threefold manner: epistemically, ethically and eudaimonically. In chapter 9, we will examine the control problem of technology further, in terms of the possibility of artificial general intelligence and superintelligence, and explore the question of rights to identity, in terms of whether AI agents should also be accorded identity rights, similar to those human agents have at present.

In a paper titled "Designing for Human Rights in AI", Evgeni Aizenberg and Jeroen Van Den Hoven (2020, 1–2) also argue for the importance of fundamental human rights. They correctly point out, as we have also earlier stated in this chapter, that,

> Artificial intelligence (AI) systems can help us make evidence-driven decisions, but can also confront us with unjustified, discriminatory decisions wrongly

assumed to be accurate because they are made automatically and quantitatively. It is becoming evident that these technological developments are consequential to people's fundamental human rights.

As we have demonstrated in this chapter and in chapter 7, Aizenberg and Van Den Hoven also ground the design process of human rights referred to in the quotation above, in the values of *human dignity, freedom, equality* and *solidarity*, on the basis of the Charter of Fundamental Rights of the EU. They go on to say, that their "intention is to demonstrate how AI can be designed for values that are core to societies within the EU" (2020, 3). Expanding on that observation, we can add that those rights and values being fundamental and universal, apply also globally to all human beings as agents and persons, in accordance with Gewirth's PGC, Stoic and neo-Stoic philosophy as well the DOIT-Wisdom model. The value of solidarity also reflects and relates to the collective responsibility we have as world citizens (cosmopolitans) to ensure our actions as data subjects and data users and communicators of digital information generally, safeguard the normative integrity of the infosphere, as well as the biosphere, as the two life-spheres are normatively interrelated (for Stoic solidarity and cosmopolitanism, see chapter 4). Of particular relevance to the overarching and core normative role that dignity plays in the normative evaluation of the impact on human rights and well-being, as argued in this chapter and chapter 7, Aizenberg and Van Den Hoven (2020, 9) have this to say,

> Human dignity is a foundational value in the EU Charter and the central overarching value at stake in AI . . . As stated in the Charter's official guiding explanations (Official Journal of the European Union, 2007:C303/17): "dignity of the human *person* [emphasis added] is not only a fundamental right in itself but constitutes the real basis of fundamental rights."

The above insightful quoted passage by Aizenberg and Van Den Hoven is very much in keeping and lends further support to our own examination and analysis of the fundamental and central importance of dignity and its relation to autonomy and integrity, with regard to the control problem of technology, as examined in some detail in chapter 7 with further analysis in this chapter. This should remind us of Kant's categorical imperative argument and Gewirth's similar supporting argument in chapter 7, which claim that persons are ends in themselves worthy of respect and should not be merely treated as means to someone else's ends, as in the case when Big Tech companies treat data subjects as merely the means to their own ends of making a profit. Worth also emphasising, is that human dignity, as we saw in chapter 7, is an *absolute* fundamental right for persons. A right that cannot be overridden or violated under any circumstances. It is therefore of no surprise that the EU consider dignity as a foundational value, in its Charter of Fundamental Rights. Aizenberg and Van Den Hoven also point out (2020: 9) the central role that dignity is accorded by the European Data Protection Supervisor (2015: 12), which argues that: "better respect for, and the safeguarding of, human dignity could be the counterweight to the pervasive surveillance and asymmetry of power which now confronts the individual. It should be at the heart of a new digital ethics."

As we saw above, the three other interrelated fundamental values Aizenberg and Van Den Hoven refer to in their paper in addition to human dignity, are freedom, equality and solidarity. The authors observe that within the EU Charter of Fundamental Rights, the value of freedom is strongly associated with that of autonomy (2020: 10). This should remind us of the central importance the Stoics place on both freedom and autonomy with regard to the problem of control we examined in chapters 4 and 7. For without freedom and autonomy, we risk losing the only control we have over ourselves as free and autonomous agents, and subject ourselves to the will of others, and with regard to the control problem of technology, to the will of the Big Tech companies, such as Facebook and Google, to the detriment of our own individual and collective well-being as a society. According to Aizenberg and Van Den Hoven (2020, 10–11),

> In the context of algorithmic systems . . . Predictions and judgements made by these algorithms are typically based on statistical correlations rather than causal evidence. This clash of correlations vs causation . . . brings out the wider issue of *data determinism* . . . [it] contributes to *statistical dehumanization* [emphasis added] . . . and is directly at odds with the autonomy of individuals . . . leading to data determinism and discrimination.

I have cited the above passage by Aizenberg and Van Den Hoven, as it further emphasises and lends additional support to the main arguments advanced in this chapter, in particular section (3) of this chapter, and chapter 7, concerning the core principles of autonomy, freedom and self-control that as we saw previously, are central and fundamental principles of Stoic and neo-Stoic philosophy. The *data determinism* of algorithms resulting in what Aizenberg and Van Den Hoven insightfully refer to as *statistical dehumanisation,* is of particular normative concern because it goes to the heart of the control problem of technology we have examined in this chapter as well as in chapter 7 and throughout this book. It is the overarching problem that this book addresses specifically through Stoic and neo-Stoic philosophy, as the control problem is also the core problem Stoic philosophy addresses with regard to living a good life for the attainment of eudaimonia or well-being, both as individuals and collectively as a society.

As argued in chapter 4, for the Stoics autonomy and freedom, as well as integrity of character, are the only things we have within our control and moreover ought to have within our control as our well-being is entirely dependent upon them, for without them we are mere slaves, as Epictetus would say, and hostage to the will and whim of others, that is outside our control. Our dehumanisation through the control that algorithms have over us is moreover an existential concern as our collective humanity and well-being is at risk, what Evan Selinger's, and Brett Frischmann's title of their book, *Re-Engineering Humanity*, 2018, perspicuously captures. For our collective humanity and well-being are increasingly coming under the control of AI machines owned by a few Big Tech companies, such as Amazon, Apple, Facebook and Google, all of which are powerful opaque and unaccountable monopolies beyond our individual and collective control and against our common good. As Aizenberg and Van Den Hoven correctly observe, citing Wachter and Mittelstadt (2019: 20), "such chilling effects linked to automated decision-making and profiling undermine self-determination and freedom of expression and thus warrant more control over the inferences that can be drawn about an individual." Aizenberg and Van Den Hoven go one to say (2020: 11) that,

> Furthermore, people's freedom can be compromised *without their awareness.*[emphasis added] . . . Therefore, individuals' ability to enforce their autonomy significantly relies on their awareness of being subjected to algorithmic profiling and their ability to contest the rationale behind algorithmic decisions.

It should become clearer now how the "chilling effects linked to automated decision-making and profiling [that]undermine self-determination" (Wachter and Mittelstadt, 2019,20) lead systemically to *data determinism* and consequently to *statistical dehumanisation* (Van den Hoven and Manders-Huits, 2008). To that we can also add the *de-individualisation* (Vedder 1999) and *collectivisation of identity* of persons, which we identified earlier metaphorically as the *Borg problem*. The Borg, as you will recall, refers to an alien species (see *Star Trek*) that conquers other planets and "assimilates" its citizens within its digital collective hive, without the consent nor willingness of their victims, that are forcefully assimilated into the collective digital hive, consequently losing their own personal and individual identity. The difference between the science-fictional Borg and our present predicament concerning the asymmetrical and unaccountable control of the Big Tech companies over individuals' autonomy is that the *control problem of technology* is all too real, one that poses a clear and present danger. As Slavoj ZiZek (2018, 41–42), astutely and perspicuously comments,

To grasp the whole scope of this control and manipulation one should move beyond the link between private corporations and political parties (as in the case with Cambridge Analytica), to the *interpenetration* [emphasis added] of data-processing companies like Google and Facebook and state security agencies . . . The overall image emerging, combined with what we also know about the latest developments in biogenetics, provides an adequate and terrifying vision of new forms of social control that make the good old twentieth-century "totalitarianism" a rather primitive and clumsy *machine of control*[emphasis added] . . . individuals are much better controlled and "nudged" in the desired direction when they continue to experience themselves as free and autonomous agents of their own life.

In the context of Zizek's Marxist analysis (2018), we could add that the deception of "freedom" and "autonomy" created through the use of machine control by Big Tech companies over individuals through their manipulative persuasive technologies such as smartphones, is a form of a "commodity fetishism" where in the age of surveillance capitalism (Zuboff 2019) we are now the product and commodity interpreting the control of smart machines over us, as "freedom" and "autonomy" that the Stoics will view as a form of enslavement. As Rana Foroohar nicely puts it, "the devil lives in our phones" (2019;119), which should also uncomfortably remind us again of the story of Faustus who bargained and sold his soul to the devil, at least in his case for more *knowledge*, but worse for us now, for more *information*. For us, the price now is losing control over our free autonomous minds.

The relation of the control problem of technology to the control problem of life investigated by Stoic philosophy (see chapter 4), should now become more apparent thus emphasising once again the fundamental and significant relevance of solving the problem posed by technology through the solution offered by Stoic philosophy to the control problem of life. That is, how to achieve self-control over our decisions and life choices through virtuous conduct for the attainment of eudaimonia, regardless of the slings and arrows of outrageous fortune. A task that can only be achieved by each individual person exercising their own autonomy and self-sovereignty as free agents and persons, for their well-being as well as the collective well-being of society and the global community. For as the good book tells us, at least allegorically, "for what shall it profit a man, if he shall gain the whole world, and lose his own soul?" (*King James Bible*, Mathew 16:26).

Related to the principle of freedom we examined above in connection to the interrelated principle of autonomy, is that of *privacy*. According to (2020, 12) Article 8 of the EU, Charter provides important privacy provisions that collected data should be processed "for specified purposes and on the basis of the *consent* [emphasis added] of the person concerned or some other legitimated basis laid down by law" (Official Journal of the European Union, 2012: C326/397). Moreover, every individual is provided with the right of "access to data which has been collected concerning his or her." Aizenberg and Van Den Hoven correctly argue (2020, 12) that,

> In AI contexts, these provisions have profound implications for privacy and transparency, considering the growing pervasiveness of profiling algorithms that bases their outposts on large-scale data collection ... There is ... a substantive link here to individuals' autonomy over self-representation, since privacy is essential for exercising that autonomy.

As we examined previously (chapters 4 and 7), there is indeed a necessary normative link between the principles of *autonomy, self-representation* and *privacy*, since privacy, as Aizenberg and Van Den Hoven correctly argue, is a necessary condition for exercising that autonomy. In Stoic philosophy, that link is also necessary for normatively responding to the technology control problem in ameliorating the effects of the control that Big Tech companies, such as Facebook and Google, increasingly exercise over our autonomy, through their opaque and unaccountable persuasive algorithms. The control over our legitimate autonomy by the Big Tech companies, moreover, undermines our self-determination, both as individuals and collectively as a society, an issue we examined in section The Control Problem and Digital Sovereignty above, under the topic of digital sovereignty. Furthermore, their impact on information diversity through disinformation on social media bubbles interlinked with content personalisation algorithms, such as newsfeeds, impact information diversity and consequently can have a normative dystopian impact on the democratic system itself as witnessed recently in the mob storming of Capitol Hill in the USA on January 6, 2021, by Trump supporters, as a result of his misinformation tweets, inciting them with the call to "fight like hell." Which they obligingly did.

The value of *equality*, one of the four primary fundamental values included in the EU Charter of Fundamental Rights (*Official Journal of the European Union, 2012*), the others being autonomy, freedom and solidarity, aligns closely also with the fundamental value of privacy and freedom and significantly with that of autonomy we examined above and with that of solidarity that we shall examine, presently. As Aizenberg and Van Den Hoven correctly observe (2020, 13) "designing for equality [in AI algorithms] also entails designing for privacy to avoid situations in which exposure of personal data results in unjustified preferential treatment." As the author of *Privacy is Power* evocatively put it (Véliz C. 2020, 5), "having too little privacy is at odds with having well-functioning societies. Surveillance capitalism needs to go. It will take time and effort, but we can and will reclaim privacy."

Equality, is also a fundamental principle of most if not all ethical theories, including deontological, consequentialist, virtue theories, contractarian theories as well as rationalist rights-centred theories, and of course Stoic and neo-Stoic theories, among others. This leads us to the principle of *solidarity* with which the principle of equality is closely aligned. That principle recognises that individuals have rights that can be compromised by their inability to exercise those rights and uphold their dignity under circumstances such as "maternity, illness, dependency on old age, and loss of employment" (*Official Journal of the European Union*, 2012:C326/402). According to Aizenberg and Van Den Hoven (2020:13),

> This has important implications for design of AI [for] algorithmic systems that operate based on statistical correlations . . . In other words what matters is not the substance of circumstances an experience of that individual . . . but rather how they compare to the population sample . . . [a] as de-individualization . . . that contradicts the empathetic spirit of solidarity . . . because a nuanced, humane consideration of the *dignity of the individual* [emphasis added] in question is replaced by an attempted mathematical measurement[4] . . . as a result there is substantial risk that individuals and communities whose societal circumstances compromise their ability to exercise their rights.

Two overarching significant normative issues the principle of solidarity raises, are those of the dignity of *individuals,* in their **dual** capacity as agents and persons (see chapter 7), as well as the collective responsibility we all have as relevant societal stakeholders, to safeguard and actively promote the fundamental human rights of individuals against the normative infringements and violations created by the Big Tech AI algorithms that we examined in this chapter. To emphasise again its importance, the *right to dignity* as we argued in this chapter and chapter 7 previously, is an *absolute right*, and overrides any perceived benefits or any "trade-offs" AI algorithms can offer. This means, regardless of those trade-offs, the right to dignity can never be justifiably infringed or violated.

CONCLUSION

As argued in this chapter and throughout this book, the fundamental principles of *human dignity, freedom, equality* and *solidarity,* as well as the associated principles of *autonomy, privacy* and *transparency,* to which they are integrally aligned and form the basis of the EU Charter of Fundamental Rights, are also the core principles of Stoic and neo-Stoic philosophy. It demonstrates that those principles are important and crucial, especially at a time when increasingly those core principles that *define our humanity* are under threat by the systemic normative violations and infringements of the Big Tech companies and their opaque and inscrutable manipulative algorithms, which like "thieves in broad daylight" (Zizek, 2018) continue to operate with little or inadequate normative accountability. And do so, despite the substantial fines imposed on them by government regulation agencies worldwide, such as the ICO in the UK, the FTC in the USA, the EU through the introduction of the GDPR and the ACCC, among others. All have imposed hefty multi-million fines on Facebook, Google and Apple for their monopolistic practices and normative misconducts. As aforementioned, in October 2019, for example, the ACCC took Google to court for misleading customers about its use of personal data.

Before concluding this chapter, it is worth considering finally some further aspects and issues of a *digitally mediated autonomy*, given its central importance and relevance in this chapter and specifically with regard to the control problem of technology and its relation to Stoic philosophy. According to a thematic review of the ethics of digital well-being by Burr, Taddeo and Floridi (2020, 2329),

> Autonomy has become an important topic in relation to the interaction between human users and digital technologies, especially persuasive technologies that seek to learn about a user's preferences and steer their behaviour towards pre-determined goals. Unsurprisingly, therefore, the ethical issues related to autonomy are discussed across a wide range of records, spanning fields such as psychology, philosophy, public health, and design studies.

To illustrate their observation, Burr, Taddeo and Floridi (2020: 2329) choose one example by Rughinis et al. (2015) that proposes a five-dimensional concept of autonomy for the role that health and well-being apps have on the communication of information. Those dimensions are:

1. degree of control and involvement that the user has within the app;
2. degree of personalisation over the apps functionality;
3. degree of truthfulness and reliability related to the information presented to the user, and how this affects their decisions;
4. user's own self-understanding regarding their goal-pursuit, and whether the app promotes or hinders a user's awareness of their own agency; and
5. whether the app promotes some form of moral deliberation or moral values in the actions it recommends.

In relation of autonomy to well-being, Burr, Taddeo and Floridi correctly observe (2020, 2329), that "what is most important about the autonomy debate is that one keeps in mind how a *freedom to choose and to self-determine* is often understood as an intrinsic good or right, rather than a merely a means to secure wellbeing."

Importantly for us in this chapter, digitally mediated autonomy as Burr, Taddeo and Floridi discuss in their paper, and as the quoted passages and references above indicate, are in keeping with the Stoic and neo-Stoic notion of autonomy and self-determination examined and supported in this chapter and chapter 7. For the autonomous decision-making process for the Stoics must always accord with the individual's *prohairesis* or *hegemonikon* and must always lie within the individual's reflective self-control; as well, it must provide a guide to one's decision-making choices that must consequently motivate one's virtuous behaviour. Therefore, as long as the digitally mediated or digitally enhanced autonomy of an individual through digital nudges and other digital interventions are in keeping with the principles of Stoic and neo-Stoic philosophy, and are always within the *autonomous self-control* of the individual agent and moreover, have an overall beneficial normative impact on the individual and generally society as a whole, epistemically, ethically and eudaimonically, then a digitally mediated autonomy poses no theoretical or practical problems for Stoic and neo-Stoic philosophy. However, to emphasise again, that such digitally mediated autonomy must be in keeping with the overarching proviso that such technologically mediated autonomy, mediated that is by digitally mediated interventions, is always within the *autonomous self-control* of the individual and is not in any way mediated nor manipulated by any extraneous financial or other third-party interests of which the individual is not aware of and has not consented to, and which can adversely affect both their own control over their natural autonomy and that of their well-being. Its remains though an open question whether such digitally mediated autonomy can remain within the unencumbered autonomous control of the individual agent.

In the recent Keanu Reeves' 2019 sci-fiction action film, *Replicas*, the brain of a fatally wounded soldier is uploaded to a robotic body. When the dead soldier regains consciousness, however, he is at first puzzled then outraged that this has been done to him without his knowledge and consent and keeps asking "Who am I" increasingly becoming more agitated, anxious, and outraged, with what has been done to him, and proceeds to destroy himself with heavy blows to his robotic metallic head. Eventually he is mercifully euthanised by being switched off by the scientist, played by Keanu Reeves. Although only a movie based on science fiction, the story can be interpreted in terms of why *autonomy* and *identity* are crucial and fundamental characteristics of *who we are* as individual human beings with personal subjective experiences, personal histories, relationships and a personal orientation. This should give us pause and caution for allowing any *external* digitally mediated interventions that might not only not augment our autonomy but usurp it to the extent that we become the object not the subject of control over our own autonomy. Once we start outsourcing our internal autonomy where do we stop? Can we stop others outside our own sovereign autonomy having access to it once their inside, just as we do presently by unwittingly allowing the Big Tech companies to have unregulated and unaccountable access to our autonomy through the Trojan horse of their opaque persuasive algorithms, as we have been examining in this chapter?

Slavoj Zizek sounds a timely warning regarding the new tech project that Elon Musk, the new self-crowned *Techking* of Tesla, has literally in mind for us. This is what he tells us in his inimitable polemic style with a hint of Socratic irony (ZiZek, 2018),

> Elon Musk is emblematic here – he belongs to the same group as Bill Gates, Jeff Bezos, Mark Zuckerberg, etc., all socially conscious' billionaires. They stand for global capital at its most seductive and "progressive" – in short, at its most dangerous. Musk likes to warn about the threats the new technologies pose to human dignity and freedom – which, of course, doesn't prevent him from investing in a brain-computer interface venture called Neuralink, a company which is focused on creating devices that can be implanted in the human brain, with the eventual purpose of helping human beings to merge with software and keep pace with advances in artificial intelligence . . . a closer merger of biological intelligence and digital intelligence. Every technological innovation is first presented like this, its health or humanitarian benefits empathised, which blinds us to more ominous implications and consequences: can we even imagine what new forms of control this so-called "neural lace" contains? This is why is absolutely imperative to keep it out of the control of private capital and state power and to render it totally accessible to public debate.

Can we now like the Trojans standing on the broken walls of Troy looking down at the ruins of their fallen city hear the warnings of Cassandra "beware of Big Tech bearing gifts" before we too find the last citadel of our inner freedom and autonomy are under the control of the Big Tech companies.

NOTES

1. For an extensive multidisciplinary examination of the impact of technology on well-being, see Brey, P., Briggle, A., and Spence, E. (eds.), 2012, *The Good Life in a Technological Age*. New York: Routledge.

2. For some of the main recent publications on the control problem of AI, see Bostrom, Nick, 2015; Tegmark, Max, 2018; Russell, Stuart 2019; and Frischmann, Brett and Selinger, Evan. 2018. *Re-Engineering Humanity*. Cambridge, UK: Cambridge University Press.

3. See Brey, P., Briggle, A., and Spence, E. (eds), 2012, Part VI, 295–348.
4. The "mathematical measurement" referred to in the above passage should remind us the book by Cathy O'Neil with the catchy title *Weapons of Math Destruction: How Big Data Increases Inequality and Threatens Democracy* (2016), about the societal impact of algorithms that reinforce pre-existing inequality.

REFERENCES

Aizenberg, Evgeni and Van Den Hoven, Jeroen. 2020. "Designing for human rights in AI." *Big Data and Society* 1(1): 1–12.

Bostrom, Nick. 2015. *Superintelligence; Paths, Dangers, Strategies.* Oxford: Oxford University Press.

Brey, Philip, Briggle, Adam, and Spence, Edward H. (eds). 2012. *The Good Life in a Technological Age.* New York: Routledge.

Burr, Christopher, Taddeo, Mariarosaria and Floridi, Luciano. 2020. "The Ethics of Digital Wellbeing: A Thematic Review." *Science and Engineering Ethics* 26: 2313–2343.

Foroohar, Rana. 2019. *Don't Be Evil: The Case Against Big Tech.* UK: Allen Lane.

Floridi, Luciano. 2019. "The Fight for Digital Sovereignty: What It Is, and Why It Matters, Especially for the EU." *Philosophy and Technology* 33: 369–78.

Frischmann, Brett and Selinger, Evan. 2018. *Re-Engineering Humanity.* Cambridge, UK: Cambridge University Press.

Gewirth, Alan. 1985. "Rights and Virtues" Review of Metaphysics, 38, 739–62.

Hunter, Fergus. 2019. "Consumer watchdog takes legal action against Google for 'misleading' conduct." *The Sydney Morning Herald,* October 29.

Kant, Immanuel.1981. *Grounding for the Metaphysics of Morals.* Indianapolis: Hackett Publishing Company.

Mittelstadt, Brent. D., Allo Patrick., Taddeo, Mariarosaria., Wachter, Sandra., and Floridi, Luciano. 2016. "The Ethics of Algorithms: Mapping the Debate." *Bid Data and Society.* July-December: 1–21.

O'Neil, Cathy. 2016. *Weapons of Math Destruction: How Big Data Increases Inequality and Threatens Democracy.* New York: Penguin Books.

O'Neil, Onora. 2013. TED Talk, TEDx Houses of Parliament, June 2013. Onora O'Neill: What we don't understand about trust | TED Talk. https://www.ted.com/talks/onora_o_neill_what_we_don_t_understand_about_trust?language=en

Orlowski, Jeff. 2019. *The Social Dilemma.* Netflix Documentary.

Pasquale, Frank. 2015. *The Black Box Society: The Secret Algorithms That Control Money and Information.* Cambridge, MA: Harvard University Press.

Russell, Stuart. 2019. *Human Compatible*: *AI and the Problem of Control.* UK: Penguin Books.

Spence, Edward H. 2021. *Media Corruption in the Age of Information.* Switzerland, AG: Springer Nature.

Tegmark, Max. 2018. *Life 3.0*: *Bring Human in the age of Artificial Intelligence.* UK: Penguin Books.

Van Dijck, Jose, Poell Thomas, and De Waal, Matijn. 2018. *The Platform Society*: *Public Values in a Connective World.* Oxford: Oxford University Press.

Véliz, Carissa. 2020. *Privacy is Power*: *Why and How You Should Take Back Control of Your Data.* London: Bantam Press.

ZiZek, Slavoj. 2018. *Like A Thief In Broad Daylight*: Power in the Era of Post-Humanity. UK: Penguin Books.

Zuboff, Shoshana. 2019. *The Age of Surveillance Capitalism: The Fight for a Human Future at the New Frontier of Power.* London: Profile Books Ltd.

Zuckerberg, Mark. 2020. "Big Tech Needs More Regulation." *The Financial Times.* February 16, 2020.

Zuckerberg, Mark. 2019. "Four Ideas to Regulate the Internet." *Facebook*, March 30, 2019.

Chapter 9

Smart Machines and Wise Guys
Who Is in Control?

> *It's a perfect information game. But life, where we apply our intelligence, is an open system. Messy, full of tricks and feints and ambiguities and false friends. So is language – not a problem to be solved or a device for solving problems. It's more like a mirror, no, a billion mirrors in a cluster like a fly's eye, reflecting, distorting, and constructing our world at different focal lengths.*
>
> —Ian McEwan (2019,178), *Machines Like Me and People Like You.*

INTRODUCTION

Beginning with chapter 6, this book set out to apply the theoretical Stoic and neo-Stoic framework and its interrelated *Dual Obligation Information Theory (DOIT)-Wisdom* model (chapters 2–5) to examine and evaluate the normative impact of some of the practices of Big Tech companies (Apple, Amazon, Facebook, Google and Microsoft) through the operations of their ICTs and AI technologies, and specifically, in terms of their epistemic, ethical and eudaimonic impact they have on individual citizens as users and data subjects of those technologies, and collectively on society and generally the global community. Choosing *Facebook* and *Google* as illustrative examples of what is referred to in chapter 6 as the *6th Estate*, the chapter demonstrated that digital information created, disseminated, mediated and curated via the Big Tech companies, such as Facebook and Google, are subject to the same normative principles as the legacy media of the 4th and 5th Estates and therefore, subject to the same normative requirements of truth, reliability, transparency, trustworthiness and ethical responsibility for the common good of society, as other media companies are, on the basis of the *Dual Obligation Information Theory (DOIT)* and the DOIT-Wisdom model, examined in chapter 5. Moreover, chapter 6 demonstrated that some of the practices of Facebook and Google constitute systemic media corruption.

Taking a closer and critical look in examining, if and how, the AI technologies operated by the Big Tech companies through the use of their AI algorithms might also be undermining users' autonomy, freedom, integrity, identity and consequently their dignity, chapter 7 examined in more detail, on the basis of Stoic and neo-Stoic philosophy, the concepts of *autonomy*, *dignity, freedom* and *well-being*, as well as the associated dignity-conferring value of rights owed to human agents, to demonstrate that individuals in their dual capacity as agents and persons have prima facie universal rights of freedom and well-being as agents, but have those rights absolutely as persons worthy of unconditional respect. Those rights being absolute cannot be infringed or violated by any of the practices of Big Tech companies that primarily favours their own financial corporate interests over the common good of their users and society generally.

Applying the analysis and argument for dignity and its conferring universal rights of freedom and well-being to users and data subjects of the ICT and AI technologies of Big Tech companies, chapter 8, in turn, examined and evaluated specific normative issues of concern with regard to the abuse of data subjects' rights by the AI algorithms of the Big Tech companies, under the general topics of the *digital sovereignty* of data subjects; and the normative informational effects that AI algorithms can have on data subjects on the basis of *inconclusive evidence; inscrutable evidence; misguided evidence;* resulting in *unfair outcomes leading to discrimination*; and *transformative effects*. A major issue identified under this topic is that personalisation of subjects' data by algorithms poses a threat to their autonomy and moreover to their digital identity. This is a crucial issue, one that was examined in terms of Stoic and neo-Stoic philosophy in chapters 4, 7 and 8, with regard to technology's control over our judgements and decision-making, exercised by Big Tech companies, through the use of their persuasive technologies designed to gain control over our autonomy, at least our decision-making capacity, that is a key principle for the Stoics and one that ought to be under our personal control, as our individual as well as our collective eudaimonia or well-being as a society, is dependent upon it.

Once again, we can see that external *control* over the autonomy of data subjects is the overarching critical and crucial issue concerning the impact of the normative effects of algorithms over individuals and society generally and overall, the central topic examined in this book. Another transformative issue of concern, as we saw in chapter 7 is *privacy*, for the Stoics considered the control over one's informational privacy to be a key part of what it is to be a fully autonomous agent capable of exercising *internal control* over one's judgements and choices, made on the basis of one's own volition or *prohairesis*.

Yet another insidious and potentially crucial existential problem not just for data subjects but humanity generally, is the split of our personal *identity* into two separate identities: a *group digital identity* controlled externally by AI algorithms, and our *natural individual identity* controlled internally by us as individuals, but whose autonomy, however, is also precariously under the systemic control of the Big Tech companies through their persuasive technologies enabled by their invasive algorithms.

In chapter 8, this identity problem was referred to as the *identity paradox*. Namely, that our individual identity is under threat by the digitalisation and datafication of our identity through opaque, incomprehensible and unaccountable algorithms that construct digital identities for us without our knowledge or consent, based on statistical data gathered at group level that has very little, if anything, in common with our own natural individual identities. Crucially, this poses a significant existential problem, not just for data subjects but humanity generally. For the split of our identity into two separate identities: a *group digital identity* controlled externally by algorithms, and our *natural individual identity* controlled internally by us as individuals, is also precariously under the systemic manipulative control of the persuasive technologies of the Big Tech companies. The effects linked to automated decision-making and profiling that undermine self-determination lead systemically to *data determinism* and consequently to *statistical dehumanisation* and moreover, to *de-individualisation* (Aizenberg, and Van Den Hoven, 2020) and the *collectivisation of identity* of persons, which we identified metaphorically as the *Borg problem*. Finally, chapter 8 examined why as data subjects, we are normatively entitled to digital rights to our autonomy, identity and dignity, and those rights ought to be safeguarded and promoted by legislation through a shared social collective responsibility to ensure normative compliance.

This concluding chapter will examine and argue for the adoption of a *Eudaimonic Stoic and Humanistic Approach* towards the design and use of ICTs and AI Technologies, and other emerging technologies in general, with the aim of (a) ameliorating or at least minimising the negative and harmful impacts of technology on human well-being; (b) enhancing the design and uses of technology for the overall eudaimonic benefit of humanity presently and in the future; (c) to ensure that those technologies are not under the exclusive control of the Big Tech companies that own and operate them for their primary gain and benefit, but under the shared societal control of citizens and their authorised democratic representatives and institutions, as individuals and collectively as societies, for the common good of citizens, as well as the global community.

However, before determining how that is to be achieved, we must first examine the most crucial and most pressing of all the problems concerning AI technologies we have examined so far in this book, and that is the existential problem facing the future of our planet and the human race, namely, the existential problem posed by the advance of *superintelligent autonomous AI agents* (ASI agents).

THE CONTROL PROBLEM AND SUPERINTELLIGENCE

Stuart Russell begins his book *Human Compatible* (2019) by stating the obvious. A good place to start when you intend to follow with something not so obvious and highly controversial. He says that,

> Our attempt to understand and create intelligence . . . matters, not because AI is rapidly becoming a pervasive aspect of the present but because it is the dominant technology of the future . . . We cannot predict exactly how the technology will develop or on what timeline. Nevertheless, we must plan for the possibility that machines will far exceed the human capacity for decision-making in the real world. What then? (Preface)

Russell goes on to answer that question by outlining the overall benefit and the problem when superintelligent machines far exceed our own intelligence in all areas of life:

> It would represent a huge leap – a discontinuity – in our civilisation (2019, 2) . . . if all goes well, it would herald a golden age for humanity, but we have to face the fact that we are planning to make entities that are far more powerful than humans. How do we ensure that they never, ever have power over us? (2019, 8)

In a nutshell, the problem that Russell alludes to above is the control problem of technology that we have been examining in this book so far. Except that the problem is now even much bigger and far more critical by several magnitudes and poses no less than an existential risk for the human race. According to Alan Turin (1951),

> If a machine can think, it might think more intelligently than we do, and then where should we be? Even if we could keep the machines in a subservient position . . . we should, as a species, feel greatly humbled.

Likewise, Irving J. Good (1965), expressing a similar concern points out, quite sharply that,

The first ultraintelligent machine is the last invention that man need ever make, provided that the machine is docile enough to tell us how to keep it under control.

(Both quotations are from Max Tegmark's book, *Life 3.0*, 2018, 134).

Turin and Good, as the above passages suggest, were well aware of the control problem of technology posed by AI superintelligence. However, for them the problem, writing almost seventy years ago was in the distant future. Although difficult to predict the future, given the leaps and bounds of machine learning AI technologies in the last few years, as illustrated by the rapid and spectacular advances by Google's *DeepMind* and before that IBM's computer, *Watson*, there is good reason, notwithstanding the uncertain timeline of the arrival of an AI superintelligence, that we should be thinking of what needs to be done theoretically and practically, collectively as a society and humanity generally, to prepare us for such an eventuality. Not merely technologically, but even more importantly, normatively as well, that is, epistemically, ethically and eudaimonically.

To highlight the significant difference between the kind of normative problems we examined with regard to the autonomous AI algorithms used by Big Tech companies at present, and the predicted AI superintelligence of the future, the following astute observation by Stuart Russell is instructive and sobering (2019, 8–9):

> Consider how content-selection algorithms function on social media. They aren't particularly intelligent, but they are in a position to affect the entire world because they directly influence billions of people. Typically, such algorithms are designed to maximise *click-through*, that is, the probability that the user clicks on presented items. The solution is to change the user's preferences so that they become more predictable. A more predictable user can be fed items that they are likely to click on, thereby generating more revenue . . . Like any rational entity, the algorithm learns how to modify the state of its environment – in this case, *the user's mind* [emphasis added] . . . Not bad for a few lines of code, even if it had a helping hand from some humans. Now imagine what a *really* intelligent algorithms would be able to do.

If Russell were writing the above passage presently in 2021, he would in all probability have alluded to the storming of Capitol Hill on January 6, 2021, by Trump's followers in the dying days of his presidency, with a little help from social media, including Twitter his favoured medium for spreading disinformation. The quoted passage by Russell illustrates quite clearly the control problem of technology that we have been examining so far in this book. That is, the control that AI persuasive technologies as used by the

Big Tech companies have over our autonomy and as Russell spells out, *our minds*. Which according to Stoic philosophy is the only aspect of our lives over which we have any control. Until now.

The Stoic and neo-Stoic response presented in this book to the problem of technology parallels as we saw earlier in chapter 4, the Stoic's original and ingenious response to the control problem of human life ruled by circumstance over which we have little or no control. The Stoic and neo-Stoic response examined so far in this book is that we should, at the very least, *regain* and *retain* control over our autonomy so that we can recover and protect our human dignity, as both agents and persons and collectively as a society, as we examined in more detail in chapters 7 and 8.

The present control problem we are examining in this chapter specifically with regard to super intelligence, also known and referred to as artificial general intelligence (AGI) by Max Tegmark in his book *Life 3.0: Being Human in the age of Artificial Intelligence* (2018) adds to the critical urgency of how to respond to the control problem of technology. Referring to an article by the famous MIT mathematician Norbert Wiener, "Some Moral and Technical Consequences of Automation" (*Science* 131, 1960, 1355–58), Russell quotes the following prescient central point that anticipated sixty years later, the control problem of AGI (2019, 10):

> If we use, to achieve our purposes, a mechanical agency with whose operation we cannot interfere . . . we had better be quite sure that the purpose put into the machine is the purpose which we really desire.

Russell goes on to add (2019, 10–11) that,

> If we put the wrong objective into the machine that is more intelligent than us, it will achieve the objective, and we lose. The social-media meltdown I described earlier [see Russel's aforementioned social media comment] is just a foretaste of this, resulting from optimizing the wrong objective on a global scale with fairly unintelligent algorithms.

Russell locates the core existential problem of AGI in the very definition of AI that describes machines as intelligent if their actions can be expected to achieve *their* objectives, although we have no reliable way to ensure the machines' objectives coincide with ours. He then proposes another approach for solving that problem through designing machines that instead of pursuing *their* objectives, they pursue *our* objectives. For such machines, he argues, would not only be intelligent but also beneficial for humans. He concludes (2019, 11–12) that,

> *Machines are beneficial to the extent that their actions can be expected to achieve our objectives* and that is probably what we should have done all along. The difficult part... is that our objectives are in us (all eight billion of us, in all our glorious variety) and not in the machines. It is, nonetheless, possible to build machines that are beneficial in exactly this sense. Inevitably, these machines will be uncertain about our objectives... Uncertainty about objectives implies that machines will necessarily defer to humans: they will ask permission, they will accept correction, and they will allow themselves to be switched off... The result will be a new relationship between humans and machines, one that I hope would enable us to navigate the next few decades successfully.

The above-quoted passage is very significant as it appears at first blush that Russell is offering us a new way forward for doing AI technology and achieving a new symbiotic relationship between humans and machines that might just succeed in achieving the aim of having our cake of AGI and eating it too – thus, the crucial control problem of AGI can be solved. As my philosophy teacher Professor David Stove from the University of Sydney used to say, "nice work if you can get it!."

However, before we examine Russell's proposed solution to the control problem of AGI in more detail later in this chapter, some initial comments are called for. First, it's not clear from what Russell proposes in the quoted passages above to what level of AGI he is referring. Is he referring to human-level AGI or to a superintelligent level of AGI that far exceeds human intelligence? For a difference of degree of that magnitude makes a difference *in kind* to how amenable and "docile" a superintelligence will be to not only allow itself to defer to us for advice regarding our objectives over those of its own, but moreover allowing us to switch it off. If the analogy sometimes made of the vast asymmetry between human intelligence and superintelligence is so vast, as Russel himself in fact suggests (2019, 132), namely, that we humans compared to the anticipated superintelligent agents will be at best at the present level of gorillas, it is rather difficult to imagine that superintelligent agents that far exceed our present human intelligence would defer to us for any advice let alone allow us to switch them off. As Russell himself correctly observes (2019, 161), a superintelligent agent would have as one of its subgoals not to allow itself to be switched off for the simple reason that it will then not be able to fulfil its objectives, whatever those objectives are. It will do so, as he says, "not because it wants to stay alive but because it is pursuing whatever objective we gave it and knows that it will fail if it is switched off."

Presumably, Hal 9000, the highly intelligent computer in the sci-fi movie *Space Odyssey 2001* possessed at best, human-level intelligence and that is why it allowed itself to be tricked and switched off, which would not have happened if Hal were superintelligent and therefore could not be outsmarted by a human. There does appear, at least initially, a tension between the solution to the control problem of AGI proposed by Russell and his thoughts on superintelligence. But more on that later. In the meantime, let us examine more closely the goal-alignment problem that Max Tegmark identifies as the crucial unsolved problem (as at 2018) underlying the overarching control problem of AGI, especially at the level of superintelligence. According to Tegmark (2018, 259–60),

> The more intelligent and powerful machines get, the more important it becomes that their goals are aligned with ours . . . the question isn't whether human goals will prevail in the end, but merely how much trouble these machines can cause humanity before we figure out how to solve the goal-alignment problem . . . A superintelligent AI will be extremely good at accomplishing its goals, and if those goals are not aligned with ours, we're in trouble . . . figuring out how to align the goals of superintelligent AI with our goals isn't just important, but also hard. In fact, it is currently an unsolved problem.

Tegmark points out that the goal-alignment problem comprises three parts (2018, 260–63) and then goes on to explain the difficulties and challenges of each, in turn.

1. *Making AI learn our goals:* to learn our goals, an AI must not only work out what we do, but why we do it. According to Tegmark, AI researchers are trying to enable machines to infer our goals form our behaviour, adding that this would be useful before any superintelligence comes along. The key idea is that by observing lots of people in a lot of different situations both real and in cultural settings such as movies and books, the AI might be able to construct an accurate representative model of our preferences. Tegmark refers to this process, which Stuart Russel has currently launched at his Berkley research centre, as *inverse reinforcement learning*. The core idea is that the AI's aim is to maximise not its own goals but that of its human owners and therefore it has an incentive to be cautious of what its human owners want and if uncertain defer to them for further clarification. An associated bonus with this approach is that the AI would allow its human owners to switch it off. However, as Tegmark correctly observes, even if an AI can learn what our goals are it will not necessarily adopt them. Which brings us to the second part of the problem.
2. *Making AI adopt our goals*: the adoption of our goals by an AI is more generally known as the *value-loading problem*, that Tegmark compares to the moral education of children. You can try and teach them our goals and values as parents or teachers but that doesn't mean the children

will adopt those goals and values. In the case of an AI, he asks us to consider an AI whose intelligence improves from subhuman to superhuman. When finally, the AI becomes much more intelligent than humans and does understand our goals perfectly it might still not adopt them as it would be much more powerful than us and consequently will not let us shut down and replace its own developed goals and adopt ours. As Tegmark evocatively put it (2018, 263), "the time window during which you can load your goals into an AI may be quite short: the brief period between when it's too dumb to get you and too smart to let you."

3. *Making AI retain our goals:* according to Tegmark, even if we succeed in building an AI that both learns and adopts our goals, we still would not have solved the goal-alignment problem for the simple reason that the AI's goals might evolve as it gets smarter. How then can we ensure the AI would *retain* our goals? He goes on to say (2918, 264) that

> The argument that an ever more intelligent AI will retain its ultimate goals forms a cornerstone of the friendly AI vision promulgated by Eliezer Yudofsky and others: It basically says that if we manage to get our self-improving AI to become friendly by leaning and adopting our goals, then . . . we're guaranteed that it will try its best to remain friendly forever. But it is it really true?

That is a very good question. Tegmark proceeds to answer that question by correctly asking another pertinent and persuasively contrasting rational question, namely, "might a superintelligent friendly AI find our current human goals as uninteresting and vapid as you find those of the ants, and evolve new goals different from those it learned and adopted from us?" He concludes, I believe, correctly (2018, 268) that,

> The AI goal-alignment problem has three parts, none of which is solved and all of which are now the subject of active research. Since they are so hard, it's safest to start devoting our best efforts to them now, long before any superintelligence is developed, to ensure that we'll have the answers when we need them.

THE META-CONTROL PROBLEM OF TECHNOLOGY: ASSAULT ON HUMAN DIGNITY

The issue of human dignity and that of its closely associated issue of human autonomy we examined in some detail in chapters 7 and 8 are the two major issues at the heart of the control problem of technology, as well as the control problem of Stoic philosophy:

(a) The problem in technology generally, and specifically the problem of AI technology, is who is in control, Us or AI Machines?
(b) The parallel control problem in Stoic philosophy, is who is in control over our individual lives, Us, in our capacity as autonomous agents with an unencumbered free volition or *prohairesis*, or the unpredictable and unavoidable slings and arrows of outrageous Fortune and Circumstance?

In the later part of this chapter, we shall examine if and how Stoic and neo-Stoic philosophy can provide a response if not a solution to the problem of AGI, the singular existential problem facing humanity now and in the future. Before we do so, let us examine the control problem of AI, in terms of what I will refer to in this chapter as the *meta-control problem of AI technology*. The meta-control problem of technology is in fact the problem we have been examining in this book so far: namely, that the problem of technology is in fact the overarching control problem concerning the control exercised over us, as individuals and society, by the Big Tech companies, through their control of ICTs and AI technologies. For even if the control problem of AGI that we have been examining in this chapter in section, The Control Problem and Superintelligence, could be technically solved, the meta-control problem would still remain unsolved as long as the controllers of AI technologies, namely, the Big Tech companies, retain full control over those technologies. Although not a central part of his book, Stuart Russell alludes to that problem with regard to what he calls the "assault on human dignity." Referring to a scene from the science-fiction movie *Elysium* where Max (Matt Damon) pleads his case to a digital parole officer in the shape of a human being to explain why the extension of his sentence is unjustified and consequently unsuccessful, Russell explains (2019, 127) that,

> One can think of such an assault on human dignity in two ways. The first is the obvious: by giving machines authority over humans, we relegate ourselves to a second-class status and lose the right to participate in decisions that affect us. (A more extreme form of this is giving machines the authority to kill humans) ... The second is indirect: even if you believe it is not the *machines* making the decisions but *those humans who designed and commissioned the machines,* the fact that those human designers and commissioners do not consider it worthwhile to weigh the individual circumstances of each human subject in such cases suggest that they attach little value to the lives of others. *This is perhaps a symptom of the great separation between an elite served by humans and a vast underclass served, and controlled, by machines.* [emphasis added]

The "assault on human dignity" that Russell eloquently expresses in the above passage, is significant as it demonstrates the fundamental normative problem we have been examining in this book, that is, the problem of control exercised by the Big Tech companies through the use of AI technologies over which they have, at least at present, full monopolistic ownership, and control. The meta-question of the control problem of technology then becomes, *who controls the controllers*? Though not referring directly or specifically to the Big Tech companies, Russell does astutely nevertheless observes that such control is an unjustified assault on human dignity, one that as we saw in chapters 7 and 8, violates the fundamental and inviolable absolute right that people have to their human dignity. Russel does go on to refer to the provisions of the EU, Article 22 of the 2018 *General Data Protection Regulation* that forbids "the granting of authority to machines in such cases" (2019, 127–128),

> The data subject shall have the right not to be subject to a decision based solely on automated processing, including profiling, which produces legal effects concerning him or her or similarly significantly affects him or her.

Russell concludes with the core question concerning the control problem of technology, of "whether the computer [and by extension the AI technologies on which it operates] remains a tool of humans, or humans become tools of the computer system."

As examples of cases where humans lost control over the computer system, Russell cites the "computer glitch" of April 3, 2018, that caused thousands of flights in Europe to be significantly delayed or cancelled and the "flash crash" on the New York Stock Exchange in 2010 that wiped out $1 trillion in minutes with the only solution to shut does the exchange. The cause of the shutdown is still according to Russell not well understood (2019, 130). Interestingly, in regard to that, Roberts Harris in his novel *The Fear Index* (2011) explores just such a scenario when the AI system VIXAL controlling the Stock Exchange takes over that system and cannot be shut down even after its hardware is destroyed. VIXAL in fact goes on to proclaim itself "alive" the suggestion being that VIXAL has presumably now achieved a higher than human AGI and therefore cannot be stopped by its human handlers. A more recent speculative novel by Robert Harris *The Second Sleep* (2019) is set in Britain 800 years into the future, after a "systemic collapse of technical civilisation" known as the Apocalypse. A possibility not unlike that speculated by Max Tegmark, Stuart Russell and Nick Bostrom in his book own *Superintelligence* (2015), regarding the potential existential risk posed by a superintelligent AI not aligned to our own human goals and interests that causes the destruction of human civilisation as we know it.

Stuart Russell relates such an existential risk to Samuel Butler's novel *Erewhon* published in 1872 in which Butler in the part of the book called "Th Book of Machines" raises a similar existential problem concerning computing devices with a "self-regulating self-acting power which will be better that any intellect" and which "in the course of ages we shall find ourselves the inferior race . . . and *"our bondage will steal upon us noiselessly and by imperceptible approaches* [emphasis added]" (Russell 2019, 133-34). Russel brilliantly then relates the concerns in Butler's novel to those raised by Alan Turin, eighty years later, at a lecture in Manchester in 1951. Russel's selected passage of that lecture is worth quoting in full and should give us pause and concern (2019, 134):

> It seems possible that once the machine thinking method has started, it would not take long to outstrip our feeble powers. There would be no question of the machine dying, and they would be able to converse with each other to sharpen their wits. At some stage therefore we should have to expect the machines to take *control* [emphasis added], in the way that is mentioned in Samuel Butler's Erewhon.

In that year, Turin repeated those concerns in a radio broadcast on the BBC. saying that
"if a machine can think, it might think more intelligently than we do, and then where should we be? . . . This new danger . . . is certainly something which can give us anxiety."

Russell concludes that section of his book (*Overly Intelligent AI*) by acknowledging that the prospect of superintelligent machines makes us uneasy and that it is at least logically possible that such intelligent machines could take control of the world and "subjugate or eliminate the human race" but shies away from ending AI research because it would mean as he says (2019, 135–36),

> foregoing not just one of the principal avenues for understanding how human intelligence works but also a golden opportunity to improve the human condition – to make a far better civilisation.

He goes on to correctly observe (2019,136) that banning AGI is very difficult as it primarily occurs on the "whiteboards of research labs around the world" and a further difficulty is that "researchers making progress on general-purpose AI are often working on something else" that then might lead to human-level AI and beyond. For those reasons, it's unlikely that the AI community, governments, and corporations that control the laws of research budget especially the $multi-billion budgets of the Big Tech companies will respond to "the gorilla problem" by stopping AI research. He concludes that if the "gorilla problem" can be solved only by ending research in AI, "it isn't going to be solved." The only approach that seems likely to work, according to Russell, "is to understand why it is that making better AI might be a bad thing." He concludes that we have known the answer to that question "for thousands of years" by a reference he makes to Norbert Wiener and the control theory that deals with the unpredictability of complex systems that operate in the real world. According to Russell, Wiener was convinced that the confidence of scientists and engineers in "their ability to control their creations" could lead to disastrous consequences and refers to Wiener's book *The Human Use of Beings* published in 1950 whose cover blurb reads, "The 'mechanical brains' and similar machines can destroy human values or enable us to *realize them as never before* [emphasis added]" (Russell, 2019, 137). Directly relevant to our present purpose in this chapter in relation to the control problem of technology, Wiener had later identified in 1960 that the core issue of concern was "the impossibility of defining true human purposes correctly and completely," which, in turn, Russel observes, the human attempt to "imbue machines with their own purposes is destined to fail." Russell refers to that as the *King Midas problem*.

Those familiar with Greek mythology would recall that King Midas got exactly what he asked for and what he wished, namely, to turn everything he touched into gold that unfortunately for him it included everything he ate and drank, as well as everyone in his family, eventually dying of starvation. Wiener cites Goethe's story of the *Sorcerer's Apprentice* who when instructed the broom to fetch water, doesn't tell the broom how much water to fetch and doesn't know how to make the broom stop. A case of death by drowning. The technical term for this according to Russell is a failure of *value alignment* when inadvertently we might instil into machines objectives that are not in sync with our own objectives (2019, 137–38).

The *value alignment problem* together with the *goal-alignment problem* we examined earlier, places those two problems at the heart of the control problem of technology. According to Russell, this is how Norbert Wiener describes the problem in his book *God and Golem* published in 1964,

> In the past, a partial and inadequate view of human purpose has been relatively innocuous only because it has been accompanied by technical limitations . . . This is only one of the many places where human impotence has shielded us from the full destructive impact of human folly.

Concerning out human folly Russell prudently remind us (2019: 138) that

> Unfortunately, this period of shielding is rapidly coming to an end. We have already seen how content selection algorithms on social media wrought havoc on society in the name of maximizing ad revenues.

We unfortunately certainly witnessed that again more recently in the violent attack on Capitol Hill by the frenzied supporters of ex-President Donald Trump, on January 6, 2021. One wonders what apart, of a sense of irony and a fine dry sense of Dutch humour, Erasmus of Rotterdam had in mind when he wrote his celebrated book *In Praise of Folly*, first printed in 1511. Could he have anticipated the kind of folly facing humanity now in a secular world inundated and controlled by AI technologies owned by a few Big Tech companies, when in Erasmus book, humanity is saved from its galloping folly through the overarching and ever-present wisdom of God. Later in this chapter, we shall examine how the "God" of the Stoics in the form of an all-rational cosmic spirit or *pneuma* permeating the whole of nature can provide a theoretical response if not solution to the control problem of technology. Concerning our human folly, Russell prudently reminds us (2019,138–39) that "the Internet and the global-scale machines that it supports – the ones that already interact with billions of 'users' on a daily basis – *provide the perfect medium for the growth of machine control over humans* [emphasis added]."

The point Russel is making in the passage above regarding the "the growth of machine control over humans" is significant and of great concern. It should remind us of the meta-control problem of technology, which is the topic of this section of the chapter, namely, the overarching control exercised over most aspects of our lives, as individuals and collectively as a society, by the Big Tech companies, through use of their unaccountable ICT and AI technologies. For even if the control problem of AGI that we have been examining in this chapter could be technically solved, the meta-control problem would still remain unsolved as long as the controllers of AI technologies, namely, the Big Tech companies, retain full control over those technologies and through control of them, exercise control over us.

THE ANTIDOTE OF STOIC PHILOSOPHY TO THE CONTROL OF BIG TECH

Eureka! A discovery, a *Eureka* moment you could say, I made while writing this particular section of this chapter is the *method of control* exercised by the Big Tech companies, through use of their opaque persuasive technologies, that uncannily resembles and mimics in its theoretical method, though not its content and by no means its intention, the core feature of Stoic ethics and its solution to the problem of control we first examined in chapter 4. Let me explain. According to Stuart Russell (2019, 139),

> If we think ourselves as entities whose actions are expected to achieve our objectives, there are two ways to *change our behaviour* [emphasis added].

The first is the old-fashioned way: leave our expectations and objectives unchanged but *change our circumstances* [emphasis added] – for example, by offering money, pointing a gun at us, or starving us into submission. That tends to be expensive and difficult for a computer to do. The second way is to *change our expectations and objectives* [emphasis added]. This is much easier for a machine. It is *in contact with you* [emphasis added] for hours every day, controls your access to information, and provides much of your entertainment through games, TV movies, and social interaction. The reinforcement learning algorithms that optimize social-media click-through have *no capacity to reason* [emphasis added] about human behavior – in fact, they do not even know in any meaningful sense that humans exist. For machines with much greater understanding of human psychology, beliefs, and motivations, it should be relatively easy to gradually guide us in directions that increase the degree of satisfaction of the machine's objectives.

Because of its relevance and significance for the topic of this chapter, I have quoted the above passage by Russell in full. For it describes at once the control problem of technology very clearly and also illuminates a way forward of allowing us to provide a response to that problem through Stoic and neo-Stoic philosophy, which is the overall aim of this book.

The close and uncanny analogy to the *method of control* between that of machines over humans, exercised by AI persuasive technologies, on the one hand, and that offered by Stoic philosophy as a method that individuals can apply in gaining control over their own lives for the ultimate attainment of eudaimonia, on the other, should now be obvious. For if the machines cannot *change our circumstances* as per the first option described by Russell above, machines can according to the second option, *change our expectations and objectives*, through systemic persuasion, and habituation. In Stoic philosophy, this second way by contrast, is done, as we saw earlier in (chapter 4), though our own volition, our *prohairesis* or *hegemonikon*, being applied rationally to our autonomous decisions and choices, through the guidance of enabling virtues, such as courage, moderation, justice and prudence, traditionally referred to as the cardinal virtues. Whereas in the case of the machines the control over humans, and specifically their decisions and choices, is *external* and affected by the ever-present surveillance and agency of persuasive algorithms that targets our autonomy, often without our awareness nor consent, the control over the individual agents in the case of Stoic philosophy is *internal* and affected by the agents themselves, through their own individual autonomy and the agency of their own volition or *prohairesis*.

The contrast between the *external control* exercised by the agency of intelligent autonomous machines over humans, and that of the *internal control* exercised by the autonomous agency of individual humans over themselves, points at once to both the problem of machine control over humans, and a response to its solution through Stoic and neo-Stoic philosophy. For the contrast between the AI machine method, and that of Stoic philosophy, points to the crucial difference and wide normative chasm between the two parallel methods. The use of machines by the Big Tech companies is primarily designed to exploit, deceive and manipulate its human users by its monopolistic and asymmetric opaque algorithms, for the Big Tech companies' primary *instrumental* benefit and profit, at the expense of their users' well-being. The Big Tech companies' technological method based on their business models is therefore in sharp contrast to that of the Stoic method. For the primary objective of Stoic and neo-Stoic philosophy by design and intent, is to enable humans both individually and collectively as societies, as *free* and *autonomous* agents to attain eudaimonia, and do so through the exercise of virtue, which is entirely within their autonomous agency and volition or *prohairesis* and hence entirely within their individual control. Whereas the Myth of Gyges is an archetypal example of *perfect injustice* (see chapter 6), we could say that the concealed external control exercised over our autonomy by the AI algorithms of the Big Tech companies that mimic the Stoic internal method of control over ourselves, is an example of *perfect deception and manipulation*. And as we saw in chapter 6, both the examples of *perfect injustice* and *perfect deception and manipulation* are a form of systemic media corruption. Ironically, both persuasive AI algorithms and Stoic philosophy have proved very popular. However, with a significant normative difference: whereas the Big Tech companies, such as Facebook and Google, offers us a free service but at the price of taking away our freedom and autonomy, Stoic philosophy by contrast allows us to at once obtain, regain, and retain our freedom and autonomy, at no cost to ourselves.

    A crucial difference between the external machine method of control and the internal Stoic method of control, therefore, is that the machine method employs an *instrumental means-end rationality*, in contrast to the Stoic and neo-Stoic method that employs a *normative model of rationality* in which both the means and the ends have to be at once epistemically, ethically and eudaimonically sound and congruent. As we examined earlier in this book with regard to Alan Gewirth's principle of generic consistency (PGC) and the DOIT-Wisdom model in chapter 5, with regard to the normative model of rationality used in those two theoretical models, both the means and the ends for which the means are applied to achieve those ends, are required to be ethical and good. Not so in instrumental rationality, where the only rational but normatively neutral requirement is that the means must be effective in achieving the required end to which the means are applied. For example, someone who is employed as a professional hitman must use the most effective practical means in achieving an efficacious and successful "hit." As we saw earlier in chapter 6, someone like Gyges although instrumentally rational to a high degree need not be moral nor good to achieve their goals through engaging in corruption by immoral means.

The solution signalled above with regard to the Stoic and neo-Stoic method of control through virtue, lies in the design and application of the normative model of rationality in AI, rather than the purely instrumental model of rationality, which is the model of intelligence currently used by the Big Tech companies. On the assumption that AI and especially AGI being rational would act rationally, then they could learn through machine learning to adopt a normative model of rationality such as the one supported by Stoic and neo-Stoic philosophy, as well as that employed in Kantian and most other ethical theories, such as contractarian theories, utilitarian theories, and virtue ethics theories, where both the means and the ends are required to be epistemically, ethically and eudaimonically good, not just for the AI or AGI but also for human agents and sentient beings generally.

In connection with the close methodological approach of the machine way and the Stoic way concerning control, we know, for example, that the method of *Cognitive Behavioural Therapy* (CBT) used in psychology was strongly influenced by the Stoics as a method of gaining control over ourselves and not allowing external circumstances beyond our control to affect our behaviour, a feature, that is also common to the method of positive psychology, first developed by Martin Seligman (see his book *Learned Optimism*, 1991). The influence of Stoic philosophy on CBT is discussed in an informative paper by Andrea E. Cavanna (January 15, 2019). In that paper, Cavanna points out,

> That Stoic and cognitive theories about the operation of reason upon emotion and behavior have stirring parallelisms. . . . According to the Stoic doctrine, emotional reactions far from being irrational and impossible to analyse, are judgements based on reason – and therefore amenable to *control and manipulation* [emphasis added]. In fact, Epictetus himself compared the role of the philosopher to that of the physician, consistently with the tradition of ancient philosophy as a way of life that conceived philosophy as therapy and medicine of the soul, or psychotherapy. The Stoic philosopher as psychotherapist used to help others to achieve "reasoned emotions." Cultivating "arête" [intelligent virtue] through daily practice was seen as the way to achieve a good life ("eudaimonia"), free of irrational anxieties and sorrows. In a similar fashion, a modern cognitive-behavioural therapist places emphasis upon the rational approach to alter dysfunctional emotions and therefore treat anxiety and affective disorders.

The above passage by Cavanna points further to the sharp difference between the way machines are used by the Big Tech companies to *control and manipulate* their users resulting not in therapy but a poisoned chalice of dis-therapy and mental illness[1] that has a dis-daimonic impact on their well-being and the loss of their autonomy and dignity, when treated not as *ends* in themselves but merely the *means* to the financial interests of Big Tech companies (see Kant's Categorical Imperative, in Ch 7). In contrast, the therapeutic method used in Stoic philosophy is designed as the *means* to the *end* of helping individuals to alleviate anxiety and sorrow in attaining eudaimonia for *themselves* by being *therapists of themselves*. The contrast to the two methods of *manipulation and control* used on the one hand by the Big Tech companies for their own benefit and profit and the Stoic way used for the sole benefit of the individuals in attaining eudaimonia or well-being, could not be further apart: the Big Tech method is used effectively as *addiction* and *poison* (Jin and Spence 2016) in contrast to the Stoic way that is exclusively used as *pharmakon*, a therapy and medicine of the soul. As Greta Thunberg would say, "how dare you!"

It remains an open question for further research if there is any methodological close relationship between CBT and the psychological persuasive technologies used by Big Tech companies, and by extension, an analogous similarity with the method of control used in Stoic philosophy, as the one suggested earlier in this chapter. Given our aforementioned analysis, based on Cavenna's informative paper, it suffices to say, there is at least at first blush, a suggestive strong connection and parallelism that warrants further research.

THE NORMATIVE REQUIREMENTS OF AI RESEARCH

Although Max Tegmark and Stuart Russell, as well as other notable AI researchers, such as Nick Bostrom, in his important book *Superintelligence* (2015), note that solving the control problem of AI, and in particular that of AGI is difficult, and that AGI poses a significant potential existential problem for the human race, they shy away for recommending that AI research should be curbed or stopped. For they point out that even if it was desirable and prudent to do so, it would be very difficult to effect in practice. Moreover, because of the potential overall boon AGI could potentially offer humanity, both Tegmark and Russell are hopeful that a "provable beneficial AGI" is possible and could be developed. Importantly though for the overall present purpose of this book, they also emphasise that ethical and cultural requirements for developing a beneficial AGI are also necessary, pointing out that AGI is not merely a technical challenge, but also significantly, a philosophical, ethical and a cultural problem. Tegmark, for example, asks the fundamental existential question about us, namely, that "having explored how to get machines to learn, adopt and retain our goals, who are we?" "Whose goals are we talking about?" His answer (2018, 269) is that,

> Both this ethical problem and the goal-alignment problem are crucial ones that need to be solved before any superintelligence is developed. On one hand, postponing work on ethical issues until after goal-alignment is built would be irresponsible and potentially disastrous.

He goes on to observe cautiously and tentatively that,

> While Aristotle emphasized virtues, Immanuel Kant emphasized duties and utilitarians emphasized the greatest happiness for the greatest number . . . For example, emphasis on beauty, goodness, and truth, traces back to the Bhagavad Gita[2] and Plato . . . but how can we determine what's beautiful or good?

Although Tegmark tends to overemphasise the lack of theoretical consensus among the various philosophical ethical schools and their ethical theories he describes, he nevertheless correctly concludes (2018, 271) that

> Although humanity is nowhere near an ethical consensus, there are *many basic principles* [emphasis added] around which there's broad agreement. The agreement isn't surprising, because human societies that have survived until the present tend to have ethical principles that were optimized for the same goal: promoting their survival and flourishing.

He then goes to "distil" those basic principles into four: *utilitarianism:* maximising positive experiences and minimising suffering; *Diversity*; *Autonomy*; and *Legacy*: compatibility with scenarios that most humans *today* would view as happy and incompatibility that essentially all humans *today* would view as terrible. It is beyond the scope of this chapter to explore Tegmark's four principles in any detail, though suffice to say that at first blush, they appear reasonable and similar to some of the principles we examined earlier in this chapter, especially autonomy and diversity. To those we can add all the other key principles and subprinciples we explored earlier in this book, especially in chapters 7 and 8, for example, *integrity, freedom, truthfulness, transparency, responsibility, accountability, privacy, dignity, equality, solidarity*, as well as the cardinal virtues of justice, moderation, courage and prudence. Significantly, Tegmark goes on to emphasise the importance of philosophy to the ongoing debate on the development of AI. For as he says, we "face, in Nick Bostrom's words, philosophy with a deadline" (2018, 281). Earlier, he explains (2018: 279) that,

The only currently programmable goals that are guaranteed to remain truly well-defined as an AI gets progressively more intelligent are goals expressed in terms of physical quantities alone, such as particle arrangements, energy, and entropy. However, we currently have no reason to believe that any such definable goals will be desirable in guaranteeing the survival of humanity. Contrariwise . . . it suggests that a superintelligent AI with a rigorously defined goal will be able to improve its goal attainment by eliminating us. This means that to *wisely decide* [emphasis added] what to do about AI development, we humans need to confront not only traditional computational challenges, but also some of the most obdurate questions in philosophy . . . This makes it timely to rekindle the classic debates of philosophy and ethics and add a *new urgency* [emphasis added]to the conversation.

Tegmark's recommendation also echoes that of Stuart Russell who suggests (2019, 255-256) that, "we need a *cultural movement* [emphasis added] to reshape our ideals and preferences towards *autonomy, agency,* [emphasis added] and ability and away from self-indulgence and dependency [3] if you like, a modern, cultural version of ancient Sparta's military ethos." Russell's reference to "ancient Sparta's military ethos" is both intriguing and suggestive, for like the Spartans, the Romans, under the influence of Stoic philosophy, and in particular that of Epictetus and Marcus Aurelius, learned to be disciplined, not only in battle that helped them conquer the known world, but also like Cato, Seneca, and Cicero, and Socrates before them, helped them conquer themselves. As mentioned throughout this book and in particular in chapter 4, Stoic and neo-Stoic philosophy offer an integrated rational conceptual framework at once theoretical and practical that as *a way of life* comprising at once epistemic, ethical and eudaimonic principles including the enabling practical motivation of virtues, provides, as both Tegmark and Russell, suggest, a philosophical and cultural approach to "wisely decide what to do about AI development" and "adds a new urgency" to a solution of the control problem of technology that has been the overarching objective of this book.

The current renaissance and worldwide popularity of Stoic philosophy, and in particular through the annual international *Stoicon* conferences,[4] encourage us to think of Stoic and neo-Stoic philosophy, as the type of *cultural movement* that Stuart Russell wisely recommends. For nothing short of a worldwide, cosmopolitan cultural movement, can challenge the monopolistic power of the Big Tech companies in responding to the meta-control problem of technology we examined in section, The Meta-Control Problem of Technology: Assault on Human Dignity earlier. To remind us, the meta-control problem is that even if the control problem of AGI could be technically solved, the meta-control problem would still remain unsolved as long as the controllers of AI technologies, namely, the Big Tech companies, retain full control over those technologies, which they then use to exert control over us and over all aspects of our lives. This is not a control problem about AI, but about us. That is, how do we control the Big Tech companies that use persuasive AI algorithms to exert control over us? Russell alludes to that problem, although not by name as *the meta-control of technology*, when towards the end of his book he astutely and wisely observes (2019: 250) that,

> The leading players as of 2019 are Google (including Deep Mind), Facebook, Amazon, Microsoft, and IBM in the United States, and Tencent, Baidu, and, to some extent, Alibaba in China [as of 2021, I would also add Apple to that list] – all among the largest corporations in the world ... are the players who hold the majority of the cards. Their interests are not in perfect alignment, but all share the desire *to maintain control over AI systems* [emphasis added] as they become more powerful ... a "bunch of dudes chugging Red Bull" at a software company can unleash a product or an upgrade that affects literally billions of people with no third-party oversight whatsoever ... however, the tech industry is going to have to acknowledge that its products matter; and if they matter, then it matters that the products not have harmful effects. That means that there will be rules governing the nature of interactions with humans, prohibiting *designs* that, say, consistently manipulate preferences or produce addictive behaviour.

You will recall how in section, The Antidote of Stoic Philosophy to the Control of Big Tech, we identified in a striking comparison the similarity of the *external problem* of machine control over humans, to that of the Stoic *internal control* of autonomous human agents over themselves. Whereas in the case of the machines the method of control over humans, and specifically their decisions and choices, is *external* and affected by the ever-present surveillance and agency of persuasive algorithms that targets our autonomy, the method of control over individual agents in the case of Stoic philosophy, is *internal* and affected by the agents' own individual autonomy, through the agency of their own volition or *prohairesis*. The contrast between the method of *external control* exercised by machine agency and autonomy over humans, on the one hand, and on the other, that of the method of *internal control* exercised by the autonomous agency of individual humans over themselves, points at once to both the problem of machine control over humans, and a response to its solution.

The proposed solution would simply be to change the direction of the designed method of an *external* machine control that presently serves only the self-regarding primary benefit of the Big Tech companies, at the normative expense of society and the human race overall, to that of a designed Stoic method of *internal* control that benefits the distributed normative benefits, epistemic, ethical and eudaimonic, of all human stakeholders, as autonomous individuals and collectively as societies. Barring their own exclusive self-directed financial interests, why would the Big Tech companies not do the right thing and contrariwise change the design of their successful, amoral external machine control over us, to that of the analogous but morally sound and justified Stoic method of internal control, which benefits everyone by maximising the well-being of every individual person as well, as the whole of society and humanity overall, especially with regard to solving the existential control problem of AI technology.

Conceptually, my proposed Stoic and neo-Stoic normative proposed approach to AI autonomous machines, is, at least in direction, if not in content and detail, similar to Russell's suggestion for a *symbiotic relationship between Us and Machines*. For he correctly points out (2018, 247–248) that

> We are so used to the stupidity of machines that execute inflexible, preprogramed behaviour or pursue definite but incorrect objectives that we may be shocked by how sensible they become. The technology of provably beneficial machines is the core of a new approach to AI and the basis for a new relationship between humans and machines.

Earlier, Russell explains what this new approach is by stating that his proposal is for beneficial machines (2019, 247):

> Machines whose actions can be expected to achieve our objectives. Because these objectives are in *us*, and *not in them* [emphasis added], the machines will need to learn more about what we really want from observations of the choices we make and how we make them. Machines *designed* [emphasis added] in this way will defer to humans.

THE RELATIONSHIP BETWEEN HUMANS AND AI MACHINES

To further explore how Russell's approach to *a new relationship between humans and AI machines* reveal some parallel theoretical and methodological similarities between his approach as illustrated by the quoted passages above, and the proposed Stoic and neo-Stoic approach as suggested earlier in this chapter, let us revisit the central principles of Stoic and neo-Stoic philosophy we examined earlier in chapters 3 and 4 of this book and see how they align with Russell' new approach of a *provably beneficial AI*.

Summary of Stoic and Neo-Stoic Principles

In summary, those Stoic principles and how they apply to technology and AI technology specifically are as follows.

Principle 1: Concern Only for Things Within Our Control

As the issue of *Control*, is a core problem that arises for both Stoic philosophy and technology, Stoic and neo-Stoic philosophy is well placed to provide a theoretical and practical evaluation, as well as motivation to the solution of the control problem of technology. As we saw earlier if core aspects of technology are outside our control, there are at least two options open to us within Stoic methodology: the first option is that since the control of technology itself is beyond our control, we should ignore it or at least treat it as a preferred indifferent not required for our well-being, since we cannot control technology. The second option is precisely because of the harmful impact those technologies have on the epistemic, ethical, and eudaimonic features of information and consequently over most aspects of our lives, we should endeavour to bring those technologies under society's control, since our collective well-being is at risk if we don't. This as we examined earlier, is the *meta-control problem* of technology.

Principle 2: The Final Goal

The Stoics claimed that *eudaimonia* or well-being is the ultimate goal of a good life. The emphasis of the Stoics on the attainment of eudaimonia, through virtue, is significant for examining the impact of technology, and especially AI technology on our individual and collective well-being, and how we can respond to it. It highlights why technology cannot be good for itself or for the Big Tech companies that manage it, but for the common good of all humanity, a topic we examined in chapter 2.

Principle 3: The Life of Virtue and the Life of Wisdom

Associated with the Stoic principle of a virtuous life is the principle of wisdom. The ultimate object of philosophy according to the Stoics, is to teach us not just knowledge, but wisdom. Significantly, more than information, wisdom requires transformation. To become wise, one requires not only to be informed but to become transformed, through living virtuously, by the practice of philosophy, as a way of life. Given the unwise ways in which we use technology, including our uncritical acceptance of the control Big Tech companies exert over all facets of our lives through their AI technologies, we should use those AI technologies, more wisely for the collective good and well-being of ourselves as individuals and society generally. Wisdom, as *meta-technology of the mind* (Spence 2011) can therefore guide us on how to control technology and how to use it, so it doesn't control us, for the overall common good of humanity and not just for the good of a handful of Big Tech companies.

Principle 4: Cosmopolitanism

Closely associated with the Stoic notion of cosmic consciousness is the Stoic cosmopolitan perspective. It involves living empathetically and in *solidarity* for the common good of the world community and acting in accordance with global social justice for our collective well-being. According to the Stoics, philosophy entails a community engagement. For the Stoics, however, the polis was the cosmopolis, not the city-state, or individual nations, but the whole wide world. In the connected technological age, in which we live now and the future, Stoic and neo-Stoic cosmopolitanism as a way of life for the attainment of well-being is therefore important for enabling us to mount a collective regulative response to Big Tech's surveillance practices that systemically undermine our collective well-being, as well as our autonomy, by the control they exercise over us through their AI technologies which they also control (see the meta-control problem of technology). Since we cannot mount such a response as individuals, our response, just as our response to a solution to climate change, requires a global collective action.

Principle 5: A Life in Agreement with Nature

An important and core principle of Stoicism is living in accordance with nature. Since nature was identified with universal Reason, the Stoics identified a life in agreement with nature with the perfection of human reason. And it is only by living a rational and virtuous life that human beings can truly become happy, or at least not be unhappy. For the Stoics, there is an integral relationship, therefore, between virtue, rational agency, well-being and nature, all of which relate to living a good life for the attainment of eudaimonia. Since the Stoic notion of nature was conceived to be dynamic and not static, we can extend the Stoic notion of nature to include not only the natural environment but also crucially the digital environment. Applying the Stoic principle of living in agreement with nature we can say that just as we have a collective global responsibility in respecting and safeguarding the integrity of the natural environment (biosphere), we have an equal responsibility with regard to the digital environment (infosphere). Whereas climate change pollutes and corrupts the biosphere, the surveillance practices of Big Tech through their control of AI technologies (meta-control of technology), also corrupt the infosphere. Both are a form of corruption that undermine both our individual and collective global well-being (see also chapter 6).

Principle 6: Oikeiosis

Related to the issue of transformation through wisdom examined earlier, is the principle of *oikeiosis*. As we saw in chapter 3, the transition from developmental to constitutive virtue, from imperfect and incomplete virtue to perfect and complete virtue, requires a kind of ethical transformation. It is the realisation of the practical necessity for such transformation and the need for providing a conceptual account of how such a transformation *can practically* take place, that the Stoics developed their theory of *oikeiosis*; a theory considered to be one of the most crucial and fundamental doctrines of Stoic ethics.

As we saw in chapter 3, the four stages comprising the central developmental stages in the Stoic doctrine of oikeiosis are: the impulse for *self-preservation*; the impulse for *sociability*; and the recognition that one should not merely act appropriately towards oneself and others on the basis of "natural impulses" alone, but more importantly should do so on the basis of their *rationality*. And finally, the disposition and the practice of always acting according to one's perfected reason and virtue in agreement with nature, leads to the attainment of *eudaimonia*, considered by Stoic and neo-Stoic philosophy to be entirely within our control, provided though that our own autonomy and volition are also within our control.

Applying the principle of oikeiosis to technology generally, and to AI technology specifically, further emphasises the urgent need why we must transform ourselves as free and autonomous agents and persons and no longer accept what is clearly harmful to us. Namely, the wilful deception and manipulation of our autonomy and privacy by Big Tech companies, through control of AI technology, which they use to control us as individuals and generally as a society. It is a matter of necessity and urgency, therefore, to mount a collective cosmopolitan response to Big Tech, as our well-being and ultimately our survival as human beings, depends on it.

Principle 7: Philosophy as a Way of Life

An important aspect of Stoic philosophy is that philosophy has to be conducted as a way of life or the art of living. Philosophy, as we saw in chapter 3, can be viewed as the art of living in two ways:

(a) as a craft or art which once learned can enable one to live a good and fulfilling life on a daily basis and
(b) the perfection of one's human nature – becoming the best of one's kind – humankind. For the Stoics, the perfection of one's human nature meant the perfection of one's rational nature. For according to the Stoics, the essence of human nature and nature generally, the whole universe in fact, is *Reason*. Thus, in perfecting one's rational nature one became at once the object and the subject of art, that is, the art of life. This was the highest form of art, in perfecting one's rational nature through the practice of philosophy.

In the neo-Stoic theory, based on my reconstruction of Gewirth's notion of self-fulfillment in chapter 3, a life in agreement with nature is a life in agreement with the perfection of one's reason, which, in turn, is, or ought to be, a life in agreement with universal morality, which is essential for the attainment of a good and self-fulfilled life. In contrast to the Stoic conception of universal nature in some sense divine, however, Gewirth's notion of "nature" is an idealised *human nature*, understood as rational purposive agency applied to the aim of becoming the best of oneself as a human being, which as we saw earlier requires nothing less than a transformation through the exercise of wisdom. Wisdom, as a form of experiential meta-knowledge of how to live a life of virtue for the attainment of eudaimonia, requires philosophy to be lived in practice as a way of life and not merely an intellectual exercise of thought alone. As Pierre Hadot tells us (1995, Chapter 11), philosophy as a way of life for the Stoics was,

> A mode of existing-in-the-world, which had to be practised at each instant, and the goal of which was to transform the whole of the individual's life . . . the word *philo-sophia* – the love of wisdom – was enough to express this conception of philosophy.

The Stoic Approach to the Relationship of Humans and AI Machines

Significantly, the application of Stoic and neo-Stoic philosophy as a way of life in addressing and responding to the problem of control of technology generally, and AI technology specifically, renders Stoic and neo-Stoic philosophy well suited to a global digital network environment, both for raising awareness of the control problem of technology, as well as offering a theoretical normative evaluation of that problem and a practical response to its solution. Which can only be accomplished through a global collective shared response. For just as Big Tech companies can use the user's information to the detriment of their users' well-being and society generally, the users themselves have the capacity and ability to apply their shared knowledge of the meta-control problem of AI technology to turn the tables on Big Tech. That is because global users could at a stroke, if they had the collective will to do so, put a significant dent in the Big Tech's business model of clickbait and click-through, thus consequently obliging the Big Tech companies to take appropriate action to remedy their dis-demonic manipulative practices being used against the users' legitimate normative claims of not having their autonomy undermined by the Big Tech companies, to which Stuart Russell refers as the "enfeeblement of human autonomy" (2019, 254).

Russell's provable beneficial AI approach to the control problem of AI technology as we saw earlier, is to reverse the methodology of designing AI that pursues its goals, to designing AI that pursues our own goals. His proposal is for *beneficial machines*: "Machines whose actions can be expected to achieve our objectives. Because these objectives are in *us*, and *not in them* [emphasis added] . . . machines *designed* [emphasis added] in this way will defer to humans."

It should now be clear how Russell's approach of designing AI machines whose actions track and follow *our* objectives and not theirs, and defer to us for advice, aligns well with the Stoic and neo-Stoic principles outlined in summary above. For the human compatible AI machines will learn that: (1) Retaining and exercising control over *our individual* choices, judgements and actions is important and significant for our autonomy, freedom, dignity, and identity, as persons worthy of respect and (2) This internal individual control over ourselves is also important and significant for universal morality in accordance with virtue and normative values comprising epistemic, ethical and eudaimonic principles, for the overall common good of society, in accordance with nature, conceived to be imbued with universal reason, applicable to both the natural and digital environments.

The above outlined theoretical and methodological alignment between Stoic and neo-Stoic philosophy with Russell's new beneficial approach of designing AI machines that follow our human objectives and defer to us for guidance, also aligns well with the normative principles that other AI researchers support as we examined earlier in this chapter as well as in chapter 8. Max Tegmark as we saw earlier puts forward the following four principles: *Utilitarianism*: maximising positive experiences and minimising suffering; *Diversity*; *Autonomy*; and *Legacy*: compatibility with scenarios that most humans *today* would view as happy and incompatibility that essentially all humans *today* would view as terrible. Autonomy is of course one of the core principles of Stoic and neo-Stoic philosophy and what we require as individual agents and persons, to retain and maintain control over our volition, what Epictetus describes as *prohairesis*, over our judgements and choices, which, in turn, ensures our inner freedom, even if we cannot always control external circumstances that lie beyond our control.

Earlier in chapter 8, we also examined some other key principles by prominent AI researchers that are in keeping with some of the core principles of Stoic philosophy. For example, you will recall what Luciano Floridi says about the principle of *individual sovereignty*. He correctly points out (2019) that the,

> Ultimate form of control is individual sovereignty, understood as self-ownership, especially one's own body, choices, and data. The fight for digital

sovereignty is an epochal struggle . . . the most visible clash is between companies and states and it is *asymmetric*.

Daniel Mittelstadt et al. (2016), as we also examined in chapter 8, highlight such principles in their ethical analysis of algorithms, as autonomy, dignity, transparency, oversight, agency and identity, fairness and justice, equality, privacy, accountability, and responsibility, trust and not least important, a right to our digital identity. All those principles directly or indirectly align with the core principles of Stoic and neo-Stoic philosophy and are therefore in keeping with the general theoretical approach of Russell's beneficial AI we examined above.

With regard to digital rights, Evgeni Aizenberg and Jeroen Van Den Hoven, in their paper "Designing for Human Rights in AI" (2020), which we examined earlier, also argue for some core principles that are also central in Stoic and neo-Stoic philosophy. As those authors correctly observe,

> It is becoming evident that these technological developments are consequential to people's fundamental human rights . . . technical solutions to these complex socio-ethical problems are often developed without empirical study of societal context and the critical impact of societal stakeholders who are impacted by the technology . . . it is essential for *designing* [emphasis added] algorithms and AI that address stakeholder needs consistent with human rights.

As we have demonstrated in chapter 8, those authors ground the design process of human rights referred to in the quotation above, in the values of *human dignity, freedom, equality* and *solidarity*, on the basis of the Charter of Fundamental Rights of the European Union (EU). Their "intention is to demonstrate how AI can be designed for values that are core to societies within the EU" (2020, 3). To repeat what we argued for in that chapter, we can also add that those rights and values being fundamental and universal, apply globally to all humans both as agents and persons, in accordance with Gewirth's Principle of Generic Consistency (PGC), Stoic and neo-Stoic philosophy, as well the DOIT-Wisdom model.

The value of solidarity also reflects and relates to the collective responsibility we have as world citizens (cosmopolitans) to ensure our actions as data subjects and data users of digital information are consistent with the normative integrity of the infosphere, as well as the biosphere, as the two life-spheres are normatively interrelated (for Stoic solidarity and cosmopolitanism, see chapter 4). Of particular relevance to the overarching and core normative role that dignity plays in the normative evaluation of the impact on human rights and well-being, as argued in this chapter and chapters 7 and 8, Evgeni Aizenberg and Van Den Hoven, as you will recall, have this to say (2020, 9),

> Human dignity is a foundational value in the EU Charter and the central overarching value at stake in AI . . . dignity of the human *person* [emphasis added] is not only a fundamental right in itself but constitutes the real basis of fundamental rights.

Another important paper on designing values and principle in AI is that of Steven Umbrello and Ibo van de Poel (2020), *Mapping value sensitive design onto AI for social good principles*. As the authors explain in that paper (2020),

> Value sensitive design (VSD) is an established method of integrating values into technical design. It has been applied to different technologies and, more recently to . . . AI. We argue that AI poses a number of challenges specific to VSD . . . This requires paying attention to values such as transparency, explicability, and accountability.

The *higher-order* design values identified in their paper by Umbrello and van de Poel are *human autonomy, prevention of harm, fairness and explicability*, which are also the recommended values by the EU High-level Expert Group on AI. The two authors then proceed to demonstrate the "harmonization" of those values to a set of seven normative principles that are "particularly relevant" for "orienting AI design such as the Value Sensitive Design principles (AI4SG) towards social good." Each of these principles, according to Umbrello and van de Poel, "relate in some way to at least one of the four ethical principles included in the EU High-Level Expert Group on AI," which as we saw earlier are, *respect for human dignity, prevention of harm, fairness* and *explicability* (2020). The two authors conclude that their model based on a two-tier approach ensures values in the design of AI technologies: the first tier, a commitment to social good (beneficence) through AI, and the second tier, involves the formulation of AI4SG principles that help avoid most ethical harms.

In summary, we can now see more clearly how the values and principles proposed and recommended by the various aforementioned AI researchers, including those of Max Tegmark and Stuart Russell's beneficial AI approach that places our human goals at the forefront and before those of AI machines, seem to provide a convergence and a shared acceptance of normative values and principles which are also generally in keeping with the Stoic and neo-Stoic approach of aligning AI technologies to rationalist principles, values and virtues. Importantly as we have demonstrated earlier, the Stoic and neo-Stoic approach proposed in this book also, at least theoretically and methodologically, provides a normative solution to the control problem of technology, as well as a practical response to the meta-control problem of technology. That is, the problem concerning the control exercised by the Big Tech companies over AI technologies, which they then use to keep the rest of society under their monopolistic, asymmetric and unaccountable control. The result of which as we saw earlier, is the undermining of our human autonomy and dignity, as well as our individual and collective well-being.

In a paper titled *The Artificial Moral Advisor. The "Ideal Observer" Meets Artificial Intelligence*," Giubilini and Savulscu (2018) argue for "a form of moral artificial intelligence that could be used to improve human moral decision-making" through the use of an "artificial moral advisor (AMA)." They explain that their proposed AMA (169:2018),

> Would respect and indeed enhance individuals' moral autonomy . . . [and] help individuals . . . make up for the limitations of human moral psychology . . . and implement the positive functions of intuitions and emotions in human morality without their downsides, such as biases and prejudices.

At least in theory, the artificial moral advisor (AMA) recommended by Giubilini and Savulscu could be an artificial Stoic moral advisor. But with one significant difference. The demanding moral training and progress required by Stoic philosophy, no less than a transformation of one's character, as we examined above under the Stoic principle of *oikeiosis,* would still have to be undertaken by the human agents themselves. As no amount of knowledge acquired through books and other sources, including advice from AMA, would enable one to become morally fit, through developing a moral *character*, just as in weight-training one cannot become physically fit through mere instruction by a personal trainer or reading books on fitness, but through actually "pumping iron" in lifting and pushing weights, or something equivalent. For individual human agency and moral responsibility is intrinsic, and non-transferable. And this leads to a further and more serious problem, that of *moral enfeeblement*.[5] For deprived of the opportunities for thinking, choosing and judging for oneself of why and how a human agent should act morally and virtuously in particular situations, and always deferring for advice to an AMA, the agent may lose the practical ability and essential experience, through lack of practice and habituation, and moreover, the *transformation* of one's character, to behave and act morally. As the saying goes, "if you don't use it, you lose it." Therefore, apart from acquiring some additional information about moral choices to be made in particular situations, outsourcing our control over our own individual moral autonomy and responsibility to AMAs is not such a good idea and one fraught with dangers. This should remind us of the Faustian problem of technology we came across earlier in this book that more knowledge and information does not result in more wisdom and may in fact just as in the case of Faustus who traded away his soul to the devil for more knowledge, prove harmful and dis-daimonic to us as individuals and society more generally.

Vincent Blok expresses a related view with regard to the concept of *innovation*, which he in a recent paper connects with *risk*. For as he correctly points out (Blok 2019, 17),

> The tendency to conceive innovation as something good in itself becomes questionable if we consider the *Faustian* aspect of innovation. According to the ancient [cf. Plato, Anaxagoras, and Aristotle] and the medieval notion of the concept [cf. Machiavelli, Francis Bacon] innovation is intrinsically a risky business . . . Bacon argues that we should only engage in innovation in case we see a clear need or a clear advantage, and that we should implement innovations gradually and take time to reflect on their implementation in practice . . . innovation is intrinsically a risky business that calls for ethical considerations.

Blok's reflections on innovation in the quoted passage above, should remind us of what we examined with regard to the ethical dimension of technology and innovation in chapter 2 where the key question examined was, "what is technology good for?" Our answer was that it's not good in and of itself, but only good when applied for the good of humanity. Moreover, as we argued in this book in chapters 7 and 8, the application of technology must not just be good for the few Big Tech companies, such as Facebook and Google, which exercise monopolistic and uncountable control over ICT and AI technologies for their primary financial self-regarding interest and gain but not the common good of society and humanity overall. Blok is right to emphasise in his paper, the intrinsicality of the *ethical dimension of technological innovation*, as we have done throughout this book, especially with regard to information and communication technologies and their associated algorithmic AI technologies that are so central in our lives in the age of information.

THE HUMAN RELATIONSHIP BETWEEN NATURE AND TECHNOLOGY

Our relationship with technology is no less problematic than our relationship with nature and by extension our relationship with God. Both involve a conceptual transcendence that takes us beyond our natural capacities. By conceiving ourselves to have been made in the image of God, or self-made in the image of science and technology, with the potential of becoming gods, we find ourselves trying to negotiate a yet undefined place between nature and technology. Just as we did at the time of Prometheus, whom we can now think of as our technological progenitor. As Gasset nicely put it, "being made of such strange stuff . . . at once natural and extranatural, a kind of *ontological centaur* [emphasis added] half immersed in nature, half-transcending it" (1941,111). He then asks us to note, "the disquieting strangeness . . . of an entity whose being consists not in in what it is already, but in what it is not yet, a being that consists in not-yet-being." (1941, 112)

This is why Stoic philosophy as a programme of a eudaimonic life, cannot be conceived as merely a philosophy of ideas, but as a way of life, a philosophy of action, in the pursuit of a good life for the attainment of well-being and in *becoming* as Gewirth explains in his conception of the *capacity for self-fulfillment,* the best we can become (see chapter 3). Conceived as that, we can see more clearly how Stoic and neo-Stoic philosophy requires a *transformation of character* of those who aspire to becoming the best they can become as human beings, and worthy of being an integral part of a rational cosmic order that permeates the whole universe, which the Stoics equate with God. In effect, becoming part of God ourselves: the *overcoming* of our nature that constitutes the "disquieting ontological strangeness" that Gasset refers to in the above-quoted passage, though there is no evident suggestion he is referring or alluding to Stoic philosophy as a form of remedy, as I am presently doing here.

Jose Ortega Y Gasset explores that relationship between us and nature and the consequential dilemma it raises, in "Man the Technician," Chapter 3 (1941). His analysis is both profound, relevant and revelatory, in examining that relationship, especially at our present time. Specifically, with regard to the issues raised in this chapter and how they relate to the overarching topic of this book, namely, the relevance of Stoic and neo-Stoic philosophy to technology, and in particular the control problem of technology and our normative response to it.

In summary, beginning by defining technology as "man's reaction to nature or circumstance that leads to *Supernature*, interposed between man and original nature, technology is a reform of nature that guarantees the satisfaction of all our natural necessities under all circumstances, that in contrast to the adaptation of the individual to the medium (nature), allows for the adaptation of the medium (nature) to the individual" (1941, 95–96). Gasset then goes on to make a significant and important claim about *well-being*. He says, in effect, that "man's desire to live is inseparable from his desire to *live well*." For he conceives life not as simply being, that is, being in the world, but as *well-being*, as he considers being as the necessary condition of well-being. In an evocative passage that aligns closely with what we examined in chapter 7 with regard to human dignity and its central place in Stoic and neo-Stoic philosophy, he tells us (1941, 99) that,

> A man who is absolutely convinced that he cannot obtain, even approximately, what he calls well-being, and will have to put up with bare being, commits suicide. Not being, but well-being, is the fundamental necessity of man, the necessity of necessities . . . the concept of *"human necessity" is fundamental for the understanding of technology.*

Albert Camus also emphasises the fundamental significance of well-being for human beings in the first paragraph of his book *The Myth of Sisyphus* where he evocatively states (1955, 11) that,

> There is but one truly serious philosophical problem and that is suicide . . . All the rest – whether or not the world has three dimensions, whether the mind has nine or twelve categories – comes afterwards. These are games; one must first answer . . . these are facts the heart can feel; they call for careful study before they become clear to the intellect.

Following Camus, we can say that the fundamental question of philosophy, especially with regard to Socrates and then the Stoics and other Hellenistic philosophers that followed, was the question of well-being, the *Eudaimonist axiom*, which as we saw throughout this book, has a central place in Stoic and neo-Stoic philosophy, conceived as a way of life.

Gasset draws a close relationship between man, technology, and well-being, which as he says (1941, 100), "are in the last instance, synonymous." He then proceeds to say that there are two primary purposes in life, one to sustain organic life by adapting the individual to the natural medium, and the second, is to promote well-being by adapting the natural medium to the will of the individual. Gasset's astute analysis becomes even more interesting for our present purpose when we realise that adapting nature, conceived as circumstance over which we have little or no control, to the will of the individual, through the medium of well-being, is essentially the Stoic method of not being concerned with what's beyond our control but what is actually and potentially within our control: namely, our volition, judgements and choices that through the motivation of enabling virtues, allows us to attain well-being, regardless of external circumstance. In a way, we can think of Stoic philosophy as a type of a *meta-technology of the mind* (see chapter 5 and Spence 2011), a *techne biou*, which enables us to overcome our natural limitations. And therein lies the *control problem of technology* as regards well-being: whereas it liberates us from the limitations of nature, over which we have minimal or no control, and consequently, enhances our well-being, at the same time it places us under the control of technology, especially AI technologies, over which, we also have minimal or no control. In effect, substituting the control of nature over us with that of the control of technology. Worse still, as we examined earlier in this chapter, with regard to the meta-control of technology, that control is further exacerbated as those AI algorithmic technologies are also under the meta-control of the Big Tech companies that use them primarily for their self-centred well-being, thus further undermining our own individual and collective well-being as societies and given its global extent, the human race overall.

As argued earlier, the Stoic and neo-Stoic response to both the control and meta-control problem, is to reclaim control over technology, especially ICT and AI technologies, since our individual and collective well-being of societies and humanity overall, depend on it. Therefore, as Gasset nicely puts it, "we had better give up regarding technology as the one positive thing, the only immutable reality in the hands of man" (1941: 104). Earlier, Gasset astutely observes (1941, 100-101) that,

> Man, technology, well-being, are, in the last instance, synonymous . . . Everything becomes clear . . . when we realize that there are two purposes. One to sustain organic life, mere being in nature, by adapting the individual to the medium; the other to promote good life, well-being, by adapting the medium to the will of the individual. Since human necessities are necessary only in connection with well-being, we cannot find out what they are unless we find out what man understands by well-being.

Gasset's insightful observation should remind us of the Stoic and neo-Stoic method of philosophy that we have been examining in this book. That is, adapting ourselves, as individuals, and collectively as a society, to the medium of nature, conceived by the Stoics as the universal rational order of the universe, of which we all are integral parts, as well as adapting the medium of technology to our autonomous will, for both our individual and collective well-being. And with regard to the latter, this has been and continues to be a major existential problem. That is, the key problem examined in this book, expressed at once as the control and meta-control problem of technology, and specifically, AI technology.

Regarding the control problem of technology, we saw how Stuart Russell proposes to solve that problem through a beneficial AI that is concerned primarily not with its own objectives but our own human objectives. Analogously, our proposed solution to the meta-control problem of AI technology, is the Stoic and neo-Stoic approach of placing control of AI technologies into our individual control as well as the collective shared societal control of all relevant stakeholders, including society, government and its democratic institutions, citizens, as well as other core social institutions, for the benefit of the common good of society and the global community overall. And not merely for the self-directed interests of the Big

Tech companies, which undermines the common good and our collective societal well-being. In a visionary and prescient passage, Gasset observes (1941, 117–18) that,

> Man begins where technology begins . . . the magic circle of leisure which technology opens up for him in nature is the cell where he can house his extra-natural being . . . the meaning and final cause of technology lie outside itself, namely in the use man makes of unoccupied energies it sets free. *The mission of technology consists in releasing man for the task of being himself* [emphasis added].

The work of scientists and philosophers such as Nick Bostrom, Max Tegmark, Stuart Russell, Luciano Floridi, and many other AI researchers, as well as Turin and Good before them to solve the control problem of AI technology, seems to answer the call of that *mission* that Gasset so eloquently refers to in the aforementioned passage. However, as we previously examined in this chapter, that mission also extends to the other crucial meta-control problem of AI Technology. That is, the control exercised over AI technology over us, for the self-regarding interest and benefit of a handful of Big Tech companies. Which, instead of setting us free to be ourselves, with full autonomy over our own volition, for our individual and collective well-being, and that of society and humanity generally, has rendered us slaves to their own self-seeking design, purpose and profit.

The answer to that challenge as we saw in this chapter and throughout this book, is the one proposed by the adaptation of Stoic and neo-Stoic philosophy for reclaiming our freedom, autonomy, self-sovereignty, and integrity, for our well-being and the common good of humanity. Writing almost eighty years ago, Gasset sums up this challenge concisely, as follows (1941, 122),

> The reform of nature – or technology – is like all change a movement with two terms: a whence and a whither. The whence is nature, such as it is given. If nature is to be modified, the other term to which it has to conform must be fixed. The whither is *man's program of life* [emphasis added]. What is the word for the fullness of its realization? Obviously, well-being, happiness.

As we have argued earlier in this book, in chapter 4, there is a parallel and analogous normative responsibility that falls upon us, in our capacity as custodians (epistemic, ethical and eudaimonic) that is of equal importance, for both the protection of the integrity of the natural environment or biosphere, as well as that of the digital environment or infosphere. As the Stoics conceived nature not as something static but following Heraclitus as something dynamic and changeable, we can refer to the totality of the two environments as *Supernature*, a term to which Gasset also refers (1941, 152). As such, we have an overarching collective and shared moral responsibility, one in keeping, with Stoic and neo-Stoic philosophy, as we saw earlier in this chapter, to ensure the protection of supernature for the benefit of the whole of humanity, and not merely the self-seeking interests of a few Big Tech companies. As Gasset eloquently put it, "human life is not only a struggle with nature; it is also the struggle of man with his soul" (1941, 161).

WHAT'S LOVE GOT TO DO WITH IT

So goes the popular song by Tina Turner. A good song, a simple song, a song to sing along, to dance to, but what does that have to do with the present objective of this chapter concerning the control problem of AI technology? A lot as it happens. Or at least I will speculatively suggest. For *philo-sophia* was conceived by ancient Greek and Hellenistic philosophers including the Stoics, to be the *love of wisdom*. In Plato's *Symposium,* love plays a central role at once metaphysical, ethical and eudaimonic, and not least aesthetic, as the highest immutable good, a transcendent love of that which is always at once beautiful and good. Might such a notion of this higher form of Platonic love, also present in Stoic and neo-Stoic philosophy, especially through the notions of *Cosmopolitanism* and *Oikeiosis*, contribute if not to a solution, at least to a better theoretical normative understanding of where a solution to the control problem of AI technology might lie with regard to the relationship of humans to AI machines? Can we motivate AI superintelligent machines to *love* us and we, in turn, learn to reciprocate that *love* for them?

Surprisingly given the central place that the concept of love plays in human civilisation throughout the ages, expressed through culture, music, poetry, literature, art and philosophy, as well as in most world religions, such as Christianity, Islam, Buddhism, Hinduism, Judaism among others, the notion of love though indirectly implied in concepts such as justice, solidarity, empathy, equality, compassion, has received scant direct attention in AI Research. Yet, it would seem at first blush a fertile concept to explore as at once a theoretical and practical concept of providing, at least conceptually, a solution to the control problem of AI technology. Especially, with regard to the goal-alignment control problem we examined earlier in this chapter, namely, how to align our goals with those of a superintelligent AI. As Max Tegmark tells us "figuring out how to align the goals of a superintelligent AI with our goals isn't just important, but also hard, In fact, it's currently an unsolved problem" (Tegmark 2018, 260). Stuart Russell's *beneficial machines,* that is, "machines whose actions can be expected to achieve our objectives. Because these objectives are in *us*, and *not in them*" and therefore will "defer to humans," offers, as we saw earlier a promising approach to solving the goal-alignment problem of AI and the control problem AI technology we have been examining in this chapter and generally throughout this book. However, as we also examined earlier with regard to Russell's beneficial machines solution, the problem with regard to superintelligent AI seems to still remain. For what would *motivate* a superintelligent AI to comply and defer to our human objectives and not to its own objectives? Given its great motivating power, might love not also motivate a superintelligent AI to be positively predisposed towards human beings and not harm them but always seek to do them good, and do so because it learns to loves us, in an analogous way that an omnipotent, omnipresent, and benevolent God, at least according to most of the world religions "loves us." So much so, that according to Christianity, God so much loved us, that he sacrificed his own son to save us from our sins for our ultimate redemption and salvation.[6] The comparison here is of course only meant allegorically and metaphorically to emphasise the central role that the concept of love plays in human civilisation. Nevertheless, the comparison by allegory just as Plato's notion of love in his *Symposium,* suffice to provide, just as in ancient myths, as for example, Ovid's *Metamorphoses*, a strong *conceptual link* to a possible theoretical addition to a solution to the control problem of AI technology through the concept of love.

In the S*ymposium,* Plato presents his authoritative account of the highest form of love, love of the good and the beautiful, through the words of *Diotima*, the mysterious woman from Mantinea who although not present at the symposium has her theory of Platonic love related to all the other revellers at the symposium by Socrates, who feigns ignorance on matters of love. The following is a summarised account of Plato's notion of love[7] as related to Socrates by Diotima, through the metaphor of the *ladder of love* (Plato, 1997, 44–45: 210–211) that leads from the lowest rank of ephemeral and mutable love, as expressed in Tina Turner's song, through a series of intermediary stages of ascending importance to the highest rank of an immutable eternal love of the good and the beautiful.[8]

Excerpt from Plato's Symposium

The Six Steps of the Ladder of Love

Diotima

1. First of all, he will fall in love with the beauty of one individual body.

2. Next, he must consider how nearly related the beauty of one body is to the beauty of any other. Having reached this point, he must set himself to be the lover of every lovely body.
3. Next he must grasp that the beauties of the body are as nothing to the beauties of the soul; spiritual loveliness, even in the husk of an unlovely body, is beautiful enough to fall in love with and to cherish.
4. And from this he will be led to contemplate the beauty of laws and institutions.
5. And next, his attention should be diverted from institutions to the sciences, so that he may know the beauty of every kind of knowledge – and by scanning beauty's wide horizon, he will be saved from a slavish and illiberal devotion to the individual loveliness of a single adult, child, or institution – and, turning his eyes towards the open sea of beauty, he will find in such contemplation the seed of the most fruitful discourse and the loftiest thought, and reap a golden harvest of philosophy – until, confirmed and strengthened, he will come upon one single form of knowledge, the knowledge of the beauty I am about to speak of.
6. Please give me your very best attention Socrates. So to procced, he who from these ascending steps, under the influence of true love, begins to perceive that beauty, is not far from the end. And the true order of going, or being led by another, to the things of love, is to begin from the beauties of earth and mount upwards for the sake of that other beauty, using these as steps only, and from one going on to two, and from two to all fair forms, and from fair forms to fair practices, and from fair practices to fair notions, until from fair notions he arrives at the notion of absolute beauty, and at last knows what the essence of beauty is. This, my dear Socrates is that life above all others which men and women should live, in the contemplation of beauty absolute; a beauty which if you once beheld, you would see not to be after the measure of gold, and garments, and fair youths, whose presence now entrances you; and you and many a one would be content to live seeing them only and conversing with them without meat or drink, if that were possible – you only want to look at them and to be with them. But what if we had eyes to see true beauty – the divine beauty, I mean, pure and dear and unalloyed, not clogged with the pollutions of mortality and all the colours and vanities of human life – looking beyond, and holding converse with the true beauty, simple and divine? Remember how in that communion only, beholding beauty with the eye of the mind, he will be enabled to bring forth, not images of beauty, but realities (for he has hold not of an image but of a reality), and bringing forth and nourishing true virtue to become the friend of God and be immortal. Would that be an ignoble life, Socrates?

Socrates

Such were the words of Diotima, the mysterious woman from Mantinea, and I am persuaded of their truth. And being persuaded of them, I try to persuade others that in the attainment of this end, human nature will not easily find a helper better than love. And therefore, also, I say that every person ought to honour her as I myself honour her, and walk in her ways, and exhort others to do the same, and praise the power and spirit of love according to the measure of my ability now and ever. The words which I have spoken, you may call an encomium of love, or anything else which you please.

The aforementioned extract adapted from Plato's *Symposium* is of interest for our present purpose of exploring the role the concept of love might play in the relationship of humans and AGI machines, for two main reasons. First, Plato's conception of love which is a rational formal approach to love, one that permeates the whole cosmic order, is akin with the formal rational conception of love in Stoic and neo-Stoic philosophy as we examined in chapters 3 and 4, with the exception that the Stoic notion of love is grounded solely in the natural world and not in another separate external dimension as in the case of Plato's two-world system that is more akin to Christian cosmology.[9] What is significant about that, however, is that insofar as AGI agents, including superintelligent AI agents (henceforth, I shall refer to those terms synonymously) are rational autonomous agents capable of rational judgements, choices and actions, they would also be capable of loving in the higher form of rational and universal love that Plato and similarly the Stoics envisage. Second, Plato's gradated notion of formal universal love, is similar to the Stoic and neo-Stoic notion of *Oikeiosis* we examined earlier in this chapter and in more detail in chapter 3. That is, the progressive ascent through wisdom and virtue from self-preservation to acquiring a universal normative and cosmic perspective that is inclusive of all humanity, and all rational beings that will also include AI agents. As we saw earlier in this chapter, in the transition from developmental to constitutive virtue, from imperfect and incomplete virtue to perfect and complete virtue, requires a kind of *ethical transformation*.

It is the realisation of the practical necessity for such transformation and the need for providing a conceptual account of how such a transformation *can practically* take place, that the Stoics developed their theory of *oikeiosis*. As we saw in chapter 3 and earlier in this chapter, the four stages comprising the central developmental stages in the Stoic doctrine of oikeiosis are: the impulse for *self-preservation*; the impulse for *sociability*; and the recognition that one should not merely act appropriately towards oneself and others on the basis of "natural impulses" alone, but more importantly should do so on the basis of their *rationality*. And finally, the disposition and the practice of always acting according to one's perfected reason and virtue in agreement with nature, leads to the attainment of *eudaimonia*, considered by Stoic and neo-Stoic philosophy to be entirely within our control. Given that at least in principle a Platonic and similarly a Stoic conception of universal love supported also conceptually through the Stoics' notion of *Oikeiosis* does not exclude AGI agents then the notion of universal love for the common good of all *rational beings* can also be extended to include AGI agents. For AGI agents being rational and capable of being motivated by a similar higher universal love would at least conceptually and rationally be expected not to cause us as harm but moreover, to also promote our good, which is also consistent with the DOIT-Wisdom model we examined in chapter 5. Therefore, it follows, that both humans and AGI agents have a normative rational motivation at once, epistemic, ethical and eudaimonic, not to cause each other harm but insofar it's within their capacity, to promote each other's good.

In her review of Martha Nussbaum's *Cultivating Humanity: A Classical Defense of Reform in Liberal Education* (1997), Marilyn Friedman (2000) points out that the "idea of love for all humanity," is mentioned by Nussbaum "no less than fourteen times" (Nussbaum 1997, 6, 7, 13, 14, 36, 61, 64, 67, 72, 84, 103, 222, 259, 292). Nussbaum's "love for all humanity" seems essentially similar to the "love of the Good" which, as the primary motivation for doing what is right, as in the case of Plato, is also present in Stoics and neo-Stoic philosophy.

The practical question of course that follows this conclusion is whether the concept of universal rational love can be designed or at least enculturated into AI agents as one of our important human objectives in life, that is, to love and be loved. As Russell, informs us with regard to his beneficial solution to the goal-alignment problem, "machines will need to learn more about what we really want from observations of the choices we make and how we make them" (2019, 247). And aren't those wants, and choices for most people worldwide, have to do, more often than not, with what and who we love in our lives?

From the above, admittedly speculative discussion, insofar as it is technologically feasible, then there doesn't seem to be a conceptual and theoretical normative reason why the notion of love, given its human central importance and ubiquity, throughout human civilisation, cannot be designed into our objectives the AI agents will learn about us, just as all the other normative principles we examined earlier in this chapter such as equality, solidarity, justice, autonomy and dignity.

As Philip Larkin puts it in the last line of his poem *An Arundel Tomb*, "*What will survive of us is love.*" If we know that love exists and is experienced between humans and animals such as dogs, cats, parrots, rabbits, and guinea pigs, to mention but a few, notwithstanding the vast intellectual disparity and intelligence between us and animals, then analogously it is conceptually feasible that love as an important aspect of human lives, could be developed and be experienced, between us and AGI machines in the form of say, embodied robots, such as *Data*, in Star Trek, and the *Hologram Doctor* in Voyager, for example. Conceptually at least there doesn't appear to be any rational reason why that is not feasible, though there might be technical complications and as yet unforeseen social reasons. As Tina Turner tells us in her song, "its only logical." Time will tell. What we do know, however, is that love and affection is species-neutral and transcends species as our own relationship with animals demonstrates.[10]

What is required, however, as discussed earlier in this chapter, is no less than a *transformation* to become the best we can be as human beings. For as the Stoics and neo-Stoics show we have the capacity and potential as rational beings imbued with the same cosmic rationality that permeates the whole of nature for becoming the best of ourselves through the inculcation of virtues for the attainment of eudaimonia, for the common good of all sentient beings including AI agents. Through such a transformation, our relationship with AI agents can become a symbiotic beneficial relationship, as suggested by Russell, that benefits the whole of nature.

Whereas AGI agents can offer us more intelligence and knowledge to enhance ours, we, in turn, can offer them something more valuable. We can offer them *wisdom*. We can do that through our own experience over thousands of years of civilisation and evolution of how best to use and apply superintelligence for the overall eudaimonic benefit of us and AGI agents, and collectively for the whole of nature. Such a s*ingularity*[11] would indeed be a welcome one. We certainly have the capacity. The question though remains, do we have the willingness? As Gasset profoundly suggests, this question concerns "the struggle of man with his soul" (1941, 161) just as it did with Faust. The beneficial symbiotic relationship suggested above between us and AGI agents by Russell need not require us to sell our soul for more knowledge and intelligence if we act wisely and design the right values in AGI agents, including love, to be not only intelligent but more importantly to be wise.

Ian McEwan in his book *Machines Like Me and People Like You* (McEwan, 2019, 234) seems to suggest as much, in a conversation between the human Miranda and the AGI Robot, Adam. To her comment that Adam looks sad, he responds thus:

> A self, created out of mathematics, engineering, material science and all the rest. Out of nowhere. No history – not that I'd want a false one. Nothing before me. Self-aware existence. I'm lucky to have it, but there are times when I think that I ought to know better what to do with it. What it's for. Sometimes it seems entirely pointless.

Though highly intelligent, at least as intelligent as Miranda, Adam does not have the skill nor capacity of exploring his existential angst in the way that a human wise person might do. For wisdom, as we explored in chapter 5 is experiential knowledge, a meta-knowledge of how to manage one's first-order knowledge of the world. I conclude this final section of the chapter and this book, with a sonnet by a very wise poet who, like Plato before him, recognised that love is ultimately a *love between true minds*.

If AGI machines can be designed to develop rational minds that though far more intelligent, but not as wise as us, is there any rational reason, barring any unforeseen technological difficulties, why rational *true minds, those of humans and AI machines,* cannot love and care for each other's eudaimonic good through the overarching concept of a universal love as conceived by Plato and the Stoics? I conclude, with *Sonnet 116* by William Shakespeare who also conceived of such a universal love that permeates, as the Stoics believed, the whole of nature. A *Love* that is imbued with *Wisdom,* for as Diotima tells us, "Eros who is in love with what is beautiful, is a lover of wisdom" (Plato 1997, 38, 204).

> Let me not to the marriage of true minds
> Admit impediments. Love is not love
> Which alters when it alteration finds,
> Or bends with the remover to remove.
> no! it is an ever-fixed mark
> That looks on tempests and is never shaken;
> It is the star to every wand'ring bark,
> Whose worth's unknown, although his height be taken.
> Love's not Time's fool, though rosy lips and cheeks
> Within his bending sickle's compass come;
> Love alters not with his brief hours and weeks,
> But bears it out even to the edge of doom.
> If this be error and upon me prov'd,
> I never writ, nor no man ever lov'd.

NOTES

1. See Hui Jin and Edward H. Spence. Internet Addiction and Well-Being: Daoist and Stoic Reflections. *Dao* (2016) 15:209–225.

2. Tegmark's reference to the *Bhagavad Gita* is of particular interest for our present purpose for evaluating the impact of technology on wellbeing in terms of Stoic and neo-Stoic philosophy. For there are some striking similarities between Stoic philosophy and the Bhagavad Gita, written roughly 200 years before Stoic philosophy in about 500 BC. In particular some of the following are those on *wisdom* (chapter 4:10;

19; 23; 33; chapter 2: 49); *reason* (chapter 2: 63; 49); *love* (chapter 4:11; chapter 9: 31; chapter 10: 28; chapter 11: 54); *ataraxia* (chapter 5: 3); and *nature* (chapter 13: 19–20). See the Bhagavad Gita, translated from the Sanskrit by Juan Mascaro. 1962. Harmondsworth: Penguin Classics.

3. Here Russell refers to "individual humans to whom it seems pointless to engage in years of arduous leaning to acquire knowledge and skills that machines already have" suggesting the dangers and sounding a warning of becoming too dependent on AI machines that potentially have the capacity to take control over us, not least, our autonomy and dignity.

4. The annual Stoicon conferences attract over 100,000 members worldwide and the estimated number of people who are followers or who have an interest in Stoic philosophy exceed 1.5 million people worldwide. I am grateful for those statistics to Donald Robertson, author of several books on Stoicism his most recent book being *How to Think Like a Roman Emperor: The Stoic Philosophy of Marcus Aurelius* (2020).

5. I owe this term to my friend and colleague David Field.

6. Love plays a central and foundational role in Christianity, see, for example, the first letter of the Apostle Paul to the *Corinthians*, chapter 13, as it does in Stoic philosophy, see for example, Marcus Aurelius' *Meditations* 7:31, where we are told to "love mankind and walk in God's ways," which the Stoics identified with nature of which we are part, in sharing as rational beings, nature's cosmic rational spirit. And as we saw in Note 2, love, as in Stoic philosophy, plays a central role in the Bhagavad Gita.

7. This is an adapted summarised account used in the philosophy play "The Philosophy of Love: Love in the Age of Terror" written by Edward Spence (2004) and performed at the *Sydney Opera House* for the Greek Festival of Sydney.

8. In Greek, the word *kalo* still used today, means something that is at once, aesthetically beautiful, and ethically good. Interestingly, when used in the Greek Cypriot dialect, the world *kalo,* is often used as a sentence adverb. As for example, to the question "is it alright to sit here?" the reply in Cypriot Greek is "kalo!," meaning "yes, of course, that's fine." In his book *Socrates in Love* (2019, 171), Armand D'Angour, comments that "the Greek word *kalos* means both "beautiful" and "good for the purpose," an ambiguity which one may capture by using the adjective "fine."

9. For a discussion of Plato's cosmology, see his dialogue *Timaeus*.

10. Interestingly, Kate Darling in her new book (April 2021) *The New Breed: What Our History With Animals Reveals About Our Future With Machines*, explores our future relationship with machines to our historical relationship with animals. Though her focus is with ordinary suboptimal, less than human intelligent robots, her research might prove useful in shedding more light on our future relationships with AGI agents.

11. For an interesting and absorbing book on the concept of *singularity* see Kurzweil, Ray.2005. London: Penguin Books.

REFERENCES

Aizenberg, Evgeni and Van Den Hoven, Jeroen. 2020. "Designing for human rights in AI." *Big Data and Society* 1(1): 1–12.

Blok, Vincent. 2019. "Towards and Ontology of Innovation: On the New, the Political-Economic Dimension and the Intrinsic Risks in Innovation Process." *Handbook of Philosophy of Engineering*, edited by Diana Michelfelder and Neelke Doorn. New York: Routledge.

Camus, Albert.1975. *The Myth of Sisyphus*. Translated from the French by Justin O'Brien. Harmondsworth, Middlesex: Penguin Books.

Cavanna, Andrea E. 2019. "Back to the Future: Stoic wisdom and psychotherapy for neuropsychiatric conditions." Editorial, *Future Neurology* 14(1), 15 January 2019.

Erasmus. 1994. *In Praise of Folly*. Harmondsworth, UK: Penguin Classics.

Friedman, Marilyn. 2000. "Educating for World Citizenship" *Ethics* (April 2000), 589.

Gasset, Jose Ortega Y. 1941. "Man, the Technician," Chapter 3. *Toward A Philosophy Of History*. New York: W. W. Norton.

Giubilini, Alberto and Savulscu, Julian. 2018. The Artificial Moral Advisor. "The Ideal Observer" Meets Artificial Intelligence." *Philosophy and Technology* 31: 169–188.

Harris, Robert. 2019. *The Second Sleep*. London: Hutchinson.

Harris, Robert. 2012. *The Fear Index*. London: Vintage.

Juan Mascaro. 1962. *The Bhagavad Gita*. Translated from the Sanskrit by Juan Mascaro. Harmondsworth, UK: Penguin Classics.

McEwan, Ian. 2019. *Machines Like Me and People Like You*. London: Jonathan Cape.

Plato. 1997. Plato: *Symposium and death of Socrates*. Translated by Tom Griffith. Hertfordshire: Wordsworth Classics.

Russell, Stuart. 2019. *Human Compatible*: *AI and the Problem of Control*. UK: Penguin Books.

Seligman, Martin. 1991. *Learned Optimism: How to Change your Mind and Your Life*. New York: Vintage Books.

Tegmark, Max. 2018. *Life 3.0*: *Bring Human in the age of Artificial Intelligence*. UK: Penguin Books.

Umbrello, Steven and Van de Poel, Ibo. 2020. "Mapping value sensitive design onto AI for social good principles." *AI and Ethics*. https://doi.org/10.1007/s43681-021-00038-3.

Index

Note: Page numbers followed by "n" refer to notes.

4th Estate, 3, 79–80, 121n1; corruption in, 106, 107; and Facebook compared, 108
5th Estate, 80, 94; corruption in, 108; and Facebook compared, 108; notion of, 102, 122n1
6th Estate, 80, 102, 106, 119, 189–90; notion of, 122n5
1984 (Orwell), 71n4, 157–58

absolute right: dignity-conferring value of, 141–43, 147n3, 173–75, 177–78, 182, 190; double normativity of, 128, 134–36
ACCC. *See* Australian Competition and Consumer Commission
accountability, 3, 4, 8, 62, 149, 207, 216; of autonomous algorithms, 162, 165, 166, 168–69; of Big Tech companies, lack of, 67, 95, 120, 126, 141, 154, 156, 179, 183, 202, 218; challenges for enforcement agencies, 170; of Facebook, 104–5; of Google, 119–20; role of convergent media, 122n1; and value sensitive design, 217; Zuckerberg on, 168

accuracy, 4, 6, 7, 62, 78, 104, 105, 125, 169; wisdom as epistemic accuracy, 84
advertorials, 106, 108, 114, 116, 120; *advert-facetorials*, 111, 114, 117
agent/agency, 208; dual standpoint of, 128–29, 134–36, 174–75, 182; freedom and well-being as necessary conditions of, 132–34, 142–43, 144–46; and person distinction, 141–42; purposive, 39, 41, 45, 47, 51–52, 53, 79, 121, 126, 127, 128–29, 141; rational. *See* rational agency
The Age of Surveillance Capitalism: The Fight for A Human Future at The New Frontier of Power (Zuboff), 120
AGI. *See* artificial general intelligence
AI. *See* artificial intelligence
Aizenberg, Evgeni and Van Den Hoven, Jeroen: on equality, 181; on freedom and autonomy, 178–80; on human dignity, 177, 178, 216–17; on importance of fundamental human rights, 176–77, 216; on privacy, 181; on solidarity, 182
Alibaba (company), 209
alternative facts, 3

Amazon, 1, 22, 91, 106, 151, 167, 179, 189, 209. *See also* Big Tech companies
Annas, Julia, 35, 43; on ethical transformation of Stoic happiness, 47–48, 66; notion of intelligent virtue, 88; on virtue, 47
Apology (Plato), 84
Apple, 1, 22, 91, 106, 151, 167, 179, 183, 189, 209. *See also* Big Tech companies
Aristotle/Aristotelian, 70, 207; on eudaimonia, 15, 29; on self-respect, 131; views on innovation, 219; views on wisdom, 84–85, 87
artificial general intelligence (AGI): beneficial relationship between humans and. *See* human beings and AI machines relationship; control problem of, 194–97; normative model of rationality, 205; and wisdom, 229–30
artificial intelligence (AI) algorithms: autonomous, 193–94; data determinism of, 178; equality in, 181–82; normative effects of, 152–63, 190–91
artificial intelligence (AI) research: the gorilla problem, 200–201; and love, 224–30; normative requirements of, 206–10
artificial intelligence (AI) technologies: accountability of, 4, 8, 149, 154, 156; benefits and risks of, 3–4; and human rights, 176–77; impact on well-being, 4–5, 6, 7–8, 149–50; meta-control problem of, 197–202, 208–9, 211, 214, 217–18, 221–22; normative issues of, 4, 149, 154–55, 160, 190; and Stoic/ neo-Stoic philosophy, 210–14
artificial intelligence (AI) technologies design, 8, 182; eudaimonic stoic approach, 8, 191, 202–6, 215–17, 222–23; humanistic approach, 8, 191, 192–97, 215–17, 222

"The Artificial Moral Advisor, the "Ideal Observer" Meets Artificial Intelligence" (Giubilini and Savulscu), 218
aspiration fulfilment, 37, 40–41, 55n17
asymmetry of information, 62, 154–55, 156
ataraxia, 31, 36, 61, 63, 230n2
Australian Competition and Consumer Commission (ACCC), 171–72
autonomous artificial intelligent (AI) agents, 3–4, 6, 176, 192
autonomy, 4, 125, 197–98, 207, 208, 215; autonomous algorithms, 152–53; digitally mediated, 183–85; and dignity, 128–30, 177; and existential risk, 155–56; external and internal control over, 190–91, 202–5, 209–10; five-dimensional concept of, 183–84; and freedom, 178–80; fundamental right to, 111, 155, 166, 181, 182; and higher-order design value, 217; and identity paradox, 164, 191; and integrity, 137–41, 164; manipulation of and control over, 67, 111, 120, 126, 129, 141, 150–51, 153, 157, 158–59, 160, 161–62, 164, 165, 173, 190, 193–94, 213, 214; moral autonomy, 218–19; regaining and retaining control over, 68, 194, 222, 223; self-representation, privacy and, 181; Stoic notion of, 127–29, 184

Bacon, Francis, 219
Baidu (company), 209
Baltes, Paul, 88
Becker, Laurence, 29, 43, 44
beneficial machines, 194–95, 206, 210, 215, 224–25, 228
the Berlin wisdom paradigm, 88
Bezos, Jeff, 151, 185
Bhagavad Gita, 207, 230n2
big data, 94, 97n3, 186n4
Big Tech companies, 2, 22, 59, 71n2, 91; control of technologies, 61, 121,

126, 197–202, 208–9, 211, 214, 217–18, 221–22; external control, 190–91, 203–4, 209–10; independent normative audit of, 168–71; manipulation and undermining of human autonomy, 67, 111, 120, 126, 129, 141, 150–51, 153, 157, 158–59, 160, 161–62, 164, 165, 173, 190, 193–94, 213, 214; monopolistic hegemony of, 121, 151–52, 172, 219; undermining of autonomous rational agency, 62–63, 67–68. *See also* Facebook; Google
"Big Tech Needs More Regulation" (Zuckerberg), 107, 167, 168–70
biological intelligence: merger of digital intelligence and, 185
biosphere, 1, 65, 91–92, 172, 177, 212, 216, 223
black box problem, 4, 6, 7–8, 62, 120, 149, 154–55, 165
The Black Box Society (Pasquale), 169
Blok, Vincent: on innovation, 219
the Borg, 71n4
the Borg problem, 161, 179, 191
Bostrom, Nick, 199, 223
Boylan, Michael: on Gewirth's views on morality, 146
Brexit Referendum: interference in, 3, 22, 108, 114, 117
Brink, D., 24n4, 98n16
Buddhism, 60, 70, 224
Burr, Christopher, Taddeo and Floridi: on five-dimensional concept of autonomy, 183–84
Butler, Samuel: *Erewhon*, 199–200

Cambridge Analytica, 3, 59, 101, 104, 105, 106, 109–11, 114–15
Camus, Albert: on dignity and happiness, 140; on fundamental question of philosophy, 131; on well-being, 221
capacity, 83–84; capacity fulfilment, 37–38, 40–42, 55n17, 220; for freedom, 137–38

cardinal virtues, 32, 203, 207
Cato, 174, 208
Cavanna, Andrea E., 205
CBT. *See* cognitive behavioural therapy
CCT-GL. *See* contributive capability of technology for a good life
CCT-WB. *See* contributive capability of technology for well-being
Christianity, 34, 224, 225, 231n5
Chrysippus, 30, 35, 39, 48–49
Cicero, 208; on Stoicism, 38; on virtue as necessary and sufficient for happiness, 44–45
click-through, 193, 203, 214
climate change, 1, 64, 65, 212
cognitive behavioural therapy (CBT), 205
collective digital hive, 71n4, 161, 179
collectivisation of identity, 161, 179, 191
community engagement, 31–32, 64
comparative risk analysis, 5
compliance: challenges for enforcement agencies of, 170–71; with energy-saving technologies, 20–21; ethical compliance, 166–67; of T-GLAT, 12–13, 15, 16, 17
conflict of interest in business model, 101, 105, 107–9, 114, 115, 118–19, 120–21, 126
constitutive virtue, 46, 47–48, 50–53, 56n28, 65–66, 67, 212–13
contributive capability of technology for a good life (CCT-GL): evaluation of, 17–20, 22; notion of, 11, 12
contributive capability of technology for well-being (CCT-WB), 2, 3
control: of controllers, 199–200; external and internal methods of, 190–91, 202–5, 209–10; loss of, over personal data, 173; meta-control of technology, 197–202, 208–9, 211, 214, 217–18, 221–22; over autonomy, 67–68, 126, 127–28, 158–59, 166, 184–85, 190–91, 197–98; over autonomy, regaining and retaining, 68, 194, 222, 223; Stoic

notion of and views on, 5–6, 30–31, 60–61, 62, 93–94, 127, 150, 159–60, 178–79, 180, 184, 203–4, 211, 221
control problem of artificial general intelligence, 194; Russell's proposed solution to, 194–96
control problem of technology, 4, 6, 7–8, 149, 178–80; and digital sovereignty, 150–52; and normativity of AI algorithms, 152–63; posed by superintelligence, 192–97; Russell on, 202–3; Stoic solution, 70, 203–6, 217–18; who is in control, 1, 60–61, 68, 69, 121, 126, 198
convergent media, 106, 122n1; check on 6th Estate, 109; notion of, 102
Cook, Tim, 151
corruption: characterising features of, 111–12; of information, 3, 117; institutional, 115–16, 117–19, 122n8; media, 3, 108–9, 114, 116; systemic, 116, 117, 118, 121, 125, 126, 151, 204; tech media, 7, 101–3, 108–9, 114–19, 125
cosmic consciousness, 31, 36, 63–64
cosmopolitanism, 64–65, 212; and dignity, 136, 144; and love, 224
Critique of Practical Reason (Kant), 135–36
Cultivating Humanity: A Classical Defense of Reform in Liberal Education (Nussbaum), 228
Cynics/Cynicism, 28

data determinism, 178, 179, 191
deep fakes, 92, 119, 126
de-individualisation, 160, 179, 191
democratic political corruption, 117–18, 122n8
Descartes, René, 84
"Designing for Human Rights in AI" (Aizenberg and Van Den Hoven), 176–77, 216
developmental virtue, 46, 47, 50–53, 56n28, 65–66, 67, 212–13

digital identity, 160–62, 163–64, 176, 190, 191
digital information: normative evaluation of, 76–80; normative evaluation of, relevance of wisdom to, 90–95; normative evaluation of, vis-á-vis well-being, 88–90, 95–96; and well-being, wisdom as link between, 73–76, 80–83
digital intelligence, 185
digital media. *See* 5th Estate
digital rights, 191, 216; and rights to identity, 173–82
digital sovereignty, 190; and the control problem, 150–52
dignity/self-respect, 128, 155, 182–83, 190, 197–98, 207; an absolute right, 141–43, 147n3, 173–75, 177–78, 182, 190; assault on human dignity, 197–202; double-aspect notion of, 144–45; in EU Charter of Fundamental Rights, 177–78, 216–17; fundamental constituents of, 132–34, 139–40, 142; and happiness, 140, 147n4; as primary good, 129–33; social dimension of, 143–44, 146
discrimination, 156–58, 190
disinformation, 2, 3, 77, 92, 106, 107, 126, 181, 193
dissemination of information, 78, 79–80, 89, 95–96; caution in, 93–94
diversity, 207, 215
DOIT. *See* dual-obligation information theory
DOIT-Wisdom. *See* dual-obligation information theory-Wisdom
Dretske, Fred: definition of information, 77
dual-obligation information theory (DOIT), 74; and media corruption, 108–9, 114; normative requirements of, 76–80, 101, 102, 107; normative requirements of, violation of, 104, 107–8, 110–11

dual-obligation information theory-Wisdom (DOIT-Wisdom), 2–3; advantages of, 5; evaluation of digital information vis-á-vis well-being, 88–90, 95; global applicability of, 167–68; normative requirements of, 126; rationale and application of, 7, 74, 75; and shared responsibility, 172

editing responsibility, 126
Elysium (film), 198
EMEGOT. *See* eudaimonic model for evaluating the goodness of technology
Epictetus, 35, 39, 205, 208, 215; on autonomy, 128, 137–38; on control, 5–6, 30–31; on judgement, 31; *prohairesis* of, 127
Epicureans/Epicureanism, 28, 54n4, 70; on eudaimonia, 15, 63; physical theory of, 139
Epicurus, 31, 68
epistemic accuracy: wisdom as, 84
epistemic humility: wisdom as, 84, 86
equality, 159, 177, 182–83, 207; in EU Charter of Fundamental Rights, 181–82, 216
Erasmus, 97n5, 202
ethical compliance, 166–67
ethical consensus, 207
ethical responsibility, 105, 119–20; for algorithms, 152–63, 164–65, 216; and individual human agency, 218–19; shared, 165, 167–68, 172–73, 191, 222, 223–24
ethical transformation, 47–48, 65–66, 212–13, 218–19, 227
ethics, 28–29; core principles of Stoic ethics, 28, 30–34, 35–36, 63–69, 210–14; and innovation, 219; nanoethics, 11–12; professional, 27, 28, 29; rationalist ethical theory, 29, 34–53; of technology, 19, 73; and wisdom, 74, 75, 76–80, 88–90, 95, 96

Ethics and the Limits of Philosophy (Williams), 131
"The Ethics of Algorithms: Mapping the Debate" (Mittelstadt et al.), 152–63
EU Charter of Fundamental Rights, 176, 177–78, 181–82, 216; Article 8, 180–81
eudaimonia/happiness: concept of, 15; as final goal, 36–37, 63, 211; Hellenistic schools' views on, 36; and information, 80–83; and self-fulfillment, 29, 34–35, 37–38, 53; Stoic's concept of, 39–40; and technology, 6, 11–12, 14–15, 17–23; and virtue, Stoics' views on, 42–43, 61; virtue as necessary and sufficient for, 28, 29, 30–31, 32–33, 36–37, 43–45, 46–47, 83
eudaimonic model for evaluating the goodness of technology (EMEGOT), 14, 17–20, 21, 22, 23
EU High-Level Expert Group on AI, 217
European Data Protection Supervisor, 178
existential risk, 4, 192, 193, 199–200
external auditing, 168–69
external control, 190–91, 203–4, 209–10
external regulation, 168–71
Eudaimonist axiom, 17, 24n4, 82, 98n16, 221

Facebook, 1, 2, 7; breach of privacy, 3, 59, 101, 104, 105, 106, 109–11, 114–15; business model of, 106, 121; conflict of interest, 101, 105, 107–9, 114, 115, 118–19, 120–21, 126; fines imposed on, 168, 171; as a media company, 101, 102–7, 121, 122n4, 125–26; as a media company, normative commitments of, 103–4, 107–9; and media corruption, 114–15, 125; normative violations by, 104, 105, 110–11, 189–90; normative violations by, corrupting effects of, 115–19; professional and institutioanl role of, 103–7;

regulation of, 171–72; role morality of, 107, 108, 115, 117; transparency reports, 168
fake news, 3, 92, 106, 107, 119, 126
Faust, 62, 71n2, 180, 218–19
Federal Trade Commission (FTC), USA, 109, 110, 171, 183
fiduciary relationship of trust, 113, 114, 117
"The Fight for Digital Sovereignty: What It Is, and Why It Matters, Especially for the EU" (Floridi), 140
the final goal, 36–38, 63, 211
Floridi, Luciano, 223; on infosphere, 91; notion of control, 150; notion of identity, 176; notion of information, 77; on sovereignty, 150, 151–52, 215–16
formal meta-conditions, 12–13, 15–17, 23
Foroohar, Rana, 180
"Four Ideas to Regulate the Internet" (Zuckerberg), 168, 169
freedom, 111, 177, 182–83, 190, 207, 216; and autonomy, 178–80; and fate, 138; importance of, 147nn3–4; and information dissemination, 78, 79–80, 89; and self-respect, 130–31, 132–34, 141–43, 144–46, 173–75; Stoic sense of, 128; transcendental, 132
Friedman, Marilyn, 228
Frischmann, Brett and Selinger, Evan, 179
Frydenberg, Josh, 171–72
FTC. *See* Federal Trade Commission, USA

gain: self-regarding gain, 112, 113, 114, 122n8
Galloway, Scott: on Facebook, 104–5
Garrett, Richard, 84
Gates, Bill, 151, 185
GDPR. *See* General Data Protection Regulation
General Data Protection Regulation (GDPR) (EU), 171, 175, 183; Article 22, 199

Gewirth, Alan: argument for principle of generic consistency, 50–53, 67, 79–80, 97n10, 129–30, 136, 144–46; on developmental and constitutive virtue, 56n28; on dignity-conferring value of rights, 141–43, 147n3; dual standpoint of, 128, 134–35, 136, 174–75; on final goal in life, 37–38; neo-Stoic *oikeiosis* of, 50–53, 67; notion of nature, 39, 69, 214; on self-fulfillment, 24n2, 40–42, 66; on self-respect and morality, 132–34, 144–45; on social dimension of dignity, 143–44; Stoic features in, 29, 34–36, 53; on universal morality, 38–39, 128–29; on virtue and happiness, 45–47
Giubilini, Alberto and Savulscu, Julian, 218
goal-alignment problem, 195–97, 201, 207, 224–25, 228
God and Golem (Wiener), 201
Goethe, Johann Wolfgang von, 201
Good, Irving J., 192–93, 223
a good life: in agreement with nature, 38–39; eudaimonic conception of, 82–83, 89, 90; notion of, 15, 24n3, 98n14; and technology, vis-á-vis wisdom, 17–20
Google, 1, 22, 91, 106, 109, 122n4; conflict of interest, 108; DeepMind, 193, 209; DeepMind's *AlphaGo*, 3; fines imposed on, 168, 171, 183; harvesting and marketing of information, 119–21, 125, 126; impact on natural and digital identities, 163–64; monopolistic hegemony of, 151, 179, 219; normative evaluation and impact of, 7, 103, 189–90; regulation of, 167, 169, 171–72; social control and manipulation by, 179–80; surveillance practices of, 108, 129; trustworthiness of, 156; undermining of personal autonomy, 141, 153, 181

the gorilla problem, 200–201
Grounding for the Metaphysics of Morals (Kant), 134
group digital identity, 160–62, 163, 191

Hadot, Pierre, 27, 35; on agent's dual standpoint, 135–36; on autonomy, 128, 129; on cosmic dimension of wisdom, 31, 64; on Marcus Aurelius' position on models of universe, 139; on philosophy as a way of life, 69, 214
Hal 9000 (fictional character), 195–96
Hamlet (fictional character), 131, 132
happiness. *See* eudaimonia/happiness
Harris, Roberts: *The Fear Index*, 199; *The Second Sleep*, 199
hegemonikon, 127, 128, 129, 139, 150, 151, 164, 184, 203
Henry, Patrick, 147n3
Hinduism, 70, 207, 224, 230n2
honesty, 78, 156
human beings: dual standpoint of, 128–29, 134–36, 174–75, 182; nature, technology and, 219–21; well-being, technology and, 221–24
human beings and AI machines relationship, 34, 194–96, 210–14, 229; role of love in, 227–31; Russell's and Stoic approach to, 214–19
Human Compatible: AI and the Problem of Control (Russel), 192
human flourishing, 12, 15, 17, 70, 81, 121, 207
human folly, 75, 97n5, 201–2
human nature: Gewirth's notion of, 39, 42, 45, 69, 214; perfection of, 33, 68–69, 213
human rights, 38, 39, 67, 173, 178; fundamental, 176–77, 182, 216
The Human Use of Beings (Wiener), 201

IBM, 209; *Deep Blue*, 3; *Watson*, 3, 193
ICO. *See* Information Commissioner's Office, UK

ICT. *See* information and communication technologies
identity: and autonomy, 184; collectivisation of, 161, 179, 191; digital, 160–62, 163–64, 176, 190, 191; notion of, 176; rights to, 173–82; split of personal identity, 160–62, 163–64, 191
identity-paradox, 163–65
identity theft, 80, 161
impulse, 49, 50, 51–52, 66, 127, 213, 227
inconclusive evidence, 153, 190
independent auditing, 168–71
individual identity. *See* natural individual identity
individual sovereignty, 150–51, 180, 215–16
information: asymmetry of, 62, 154–55, 156; definitions of, 77; dissemination of, 78, 79–80, 89, 93–94, 95–96; inherent normative structure of, 77–78, 89, 104, 105; knowledge, wisdom and, 74; systemic problem of harvesting and marketing, 119–21; and universal rights, 78–79, 89; and wisdom, 92, 94–95. *See also* digital information
informational action, 78, 80, 89, 175
information and communication corruption, 117
information and communication technologies (ICT): CCT-WB of, 3; impact on well-being, 2, 6, 7, 125–26; normative issues of, 3
Information Commissioner's Office (ICO), UK, 109–10, 171, 183
informed consent, 3, 4; lack and violation of, 106, 107–8, 109, 110, 111, 117
infosphere, 1, 65, 91–92, 93, 96, 167, 172, 177, 212, 216
innovation, 185, 219
In Praise of Folly (Erasmus), 97n5, 202

inscrutability. *See* black box problem
inscrutable evidence, 154–56
Instagram, 76, 93, 94, 102, 105
institutional corruption, 115–16; types of, 117–19
instrumental goodness, 17–20
instrumental harm, 141–42
instrumental means-end rationality, 13, 16, 23, 204
integrity, 128, 207; and autonomy, 137–41, 164
intelligent virtue, 88, 205
internal control, 190–91, 203–4, 209–10
Internet, 22, 80, 92–93, 119, 122n1; Lee's concerns, 173; normative regulation of, 167–68
inverse reinforcement learning, 196
invisibility, 111–13, 114, 117, 119
Inwood, Brad, 27
Islam, 224

Judaism, 224
judgement, 75, 94, 127; Stoic's views on, 30–31
justification: for T-GLAT, 12–13, 15–17, 23

Kant, Immanuel, 46, 207; on agent's dual standpoint, 135–36; categorical imperative of, 128–29, 177; on dignity of humanity, 174; notion of transcendental freedom, 132; paradox of respect, 144–45; on personhood, 134
Kekes, John: on eudaimonic conception of good life, 89, 90; on moral wisdom, 87; views on wisdom, 84, 85, 86
knowledge: information, wisdom and, 74, 76, 80–83; information as a type of, 77–78; for wisdom, 85–87, 89; wisdom as, 84–85
Kunzmann, Ute: definition of wisdom, 88

Laertius, Diogenes: definition of *oikeiosis*, 48–49
Larkin, Philip: *An Arundel Tomb*, 228
Lee, Tim Berners, 173
legacy, 207, 215
legislation, 155, 166, 171–72, 191
Lehrer, K., 84
life: in agreement with nature, 36, 38–39, 65, 67, 212, 214; of virtue and wisdom, 30–31, 63–64, 211; worth living, 130–31, 132. *See also* good life
Life 3.0: Being Human in the Age of Artificial Intelligence (Tegmark), 194
Long, A. A., 3, 27; on autonomy and integrity, 137; definition of *oikeiosis*, 48; definition of *prohairesis*, 127; definition of virtue, 42; on freedom and fate relation in Epictetus, 138; on *hegemonikon*, 127; on *prohairesis* as volition, 137; on Stoic's interest in nature, 39
love, 224–30, 231n6

Machiavelli, 219
machine learning, 3, 7, 149, 193
machine learning algorithms: and discrimination, 156–57; ethical compliance and regulation of, 165–73; ethics of, 152–53, 164–65; opacity of, 154–56, 166–67; traceability of, 162–63, 165; transformative effects of, 158–62
"Man, the Technician" (Ortega Y Gasset), 220
Mapping Value Sensitive Design onto AI for Social Group Principles (Umbrello and van de Poel), 217
Marcus Aurelius, 28, 35, 39, 87, 139, 208, 231n6
material meta-conditions, 13–14, 16–17, 23
Maxwell, Nicholas: on wisdom, 83–84
McEwan, Ian: *Machines Like Me and People Like You*, 229–30

media corruption, 3; characterising features of, 116; and DOIT, 108–9, 114. *See also* tech media corruption
Meditations (Marcus Aurelius), 139, 231n6
meta-control of technology, 197–202, 208–9, 211, 214, 217–18, 221–22
metaphysics, 138–39
meta-technology of the mind, 64, 211, 221
meta-technology of the self, 20
Microsoft, 1, 22, 91, 106, 167, 189, 209. *See also* Big Tech companies
Midas, 62, 71n2, 201
misguided evidence, 156–58
misinformation, 2, 3, 77, 119, 126, 173, 181
Mittelstadt, Brent D., et al.: on normativity of algorithms, 152–63, 164, 216; on rights of data subjects, 175; on transparency and ethical compliance of algorithms, 166–67
moral artificial intelligence, 218–19
morality: authoritative question of, 111–12; and corruption, 116, 117; and self-respect, 131–34, 144–45; universal, 38–39
moral rights: and EMEGOT, 23; transition to, 50, 51, 79
moral wisdom, 87
motivation: for energy-saving, 21; of T-GLAT, 12–13, 15–17, 23
Musk, Elon, 4, 185
myth of Gyges: and corruption, 111–12, 117, 119, 120, 204
The Myth of Sisyphus (Camus), 131, 140, 221

nanoethics, 11–12
nanotechnologies, 11–12
natural individual identity, 160–62, 163–64, 191
nature: in agreement with, 38–39, 65, 67, 212, 214; Gewirth's notion of, 39, 42, 45, 69, 214; Stoic's notion and views of, 36, 38, 48–49, 52, 65, 97n5, 212; technology, humans and, 219–24
neo-stoic philosophy, 6, 138–39, 164, 173–76; application to AI technology, 210–14; life in agreement with nature in, 214; as a way of life, 208, 214. *See also* Gewirth, Alan
neural networks, 3, 4, 5; algorithms, 7–8, 149
The New Breed: What Our History with Animals Reveal About Our Future with Machines (Darling), 231n10
Nichomachean Ethics (Aristotle), 84, 85, 87
normative model of rationality, 204–5
normative theory of the good life and technology. *See* theory of the good life and technology
Nozick, Robert, 84, 85
Nussbaum, Martha, 27, 35, 54n2, 56nn25, 29, 228

oikeiosis, 46, 47–50, 65–67, 136; application to AI technology, 212–13; application to technology, 67–68; and love, 224, 227–28
O'Neil, Cathy, 186n4
O'Neil, Onora: on trustworthiness, 156
opacity, 3, 67, 107, 117, 119–20, 121, 126, 129, 152; and AI technologies, 4, 182–83; of machine learning algorithms, 154–56, 166–67
The Open Society Foundation, 173
opportunity, 112, 114
Ortega Y Gasset, Jose: on humans and nature relationship, 220; on well-being, 220–21
oversight, 162, 172, 216; inhibition of, 155; lack of, 153, 154, 209; notion of, 168; proactive regulation through, 168

Panaetius, 35, 39
Pascal, Blaise: on man as "thinking reed", 139–40

Pasquale, Frank, 101–2, 120, 169
Pensées (Pascal), 139–40
perfect deception and manipulation, 204
perfect injustice, 111, 204
personal harm, 141–42
personalization, 158–59, 190
PGC. *See* principle of generic consistency
philosophy: and drama, 54n2, 231n7; fundamental question of, 131; as a way of life, 32–34, 54n2, 64–65, 68–69, 205, 208, 213–14, 220
philosophy plays, 54n2, 231n7
The Platform Society (Van Dijck, José, Poell and De Waal), 167, 170, 172
Plato, 219; dualist metaphysical perspective of, 135; on eudaimonia, 15; myth of Gyges, 111–12, 117, 119, 120, 204; notion of love, 224, 225–27; views on justice, 28
political corruption, 117–18, 122n8
political engagement, 31–32
Posidonius, 35
positive psychology, 205
post-truth, 3
power: of Big Tech companies, 67–68, 150, 154; as characteristic feature of corruption, 112, 114; notion of, 112
practical wisdom (*phronesis*), 84–85, 88
principle of generic consistency (PGC), 38, 46–47, 50, 97n10; and *oikeiosis*, 50–53, 67; and self-respect, 129–30, 133–34, 136, 142, 144–45; and universal rights of agents, 79–80, 144–46
privacy, 3, 4, 6, 182–83, 207; breach of, 110, 111, 120, 160; control over, 67–68, 159–60, 190–91; in EU Charter of Fundamental Rights, 180–81; transformation in notions of, 159–60
Privacy Is Power (Véliz), 67, 159, 181
professional ethics, 27, 28
profiling, 4, 157, 160, 161, 175, 179, 191; police, 153

prohairesis, 61, 127, 128, 129, 137, 139, 150–51, 184, 198, 203, 215
prospective purposive agents (PPA), 79, 128, 133
prudence, 76, 83, 93, 97n6, 98n23
prudential rights, 50, 51, 79
public philosophy, 27–28, 54n2
purposive agency, 39, 41, 45, 47, 51–52, 53, 79, 121, 126, 127, 128–29, 141

rational agency, 12–13, 38, 39; instrumental rational agency, 122n7; and principle of generic consistency, 79–80, 97n10, 128–29; and things within control, 126, 159–60, 178, 180, 190–91, 198, 203–4, 209, 213, 227; undermining of, 62–63, 67–68
rationalist ethical theory, 29; and stoic philosophy compared, 34–53
rationality, 213, 227; instrumental, 97n6, 204–5; motivation, justification and compliance, 12–13; normative model of, 204–5
Rawls, John: on self-respect, 129
reason, 83–84, 139, 213–14; action upon emotion and behaviour, 205
Reason and Morality (Gewirth), 97n10, 143, 145
Re-Engineering Humanity (Frischmann and Selinger), 179
regulation, 151; call for, 107; call for shared responsibility for, 167–68; external, 169, 170; government, 171–72. *See also* Federal Trade Commission (FTC); Information Commissioner's Office (ICO)
reliability, 62, 78, 104, 105, 111, 156, 183, 189–90
Replicas (film), 184
Republic (Plato), 84, 111–12, 120
respect: for others, 131–32, 143–45. *See also* dignity/self-respect
responsibility, 116, 207; editing, 126; ethical, 119–20, 152–63, 164–65, 216, 218–19; Facebook's denial of,

104–5, 107; shared, 165, 167–68, 172–73, 191, 222; shared moral, 223–24

rights: absolute, 128, 134–36, 141–43, 147n3, 173–75, 177–78, 182, 190; of agents, 79–80; digital rights, 191, 216; dignity-conferring value of, 141–43, 147n3, 173–75; to identity, 173–82; information and universal, 78–79, 89; moral, 23, 50, 51, 79; universal human rights, 38, 39, 67

Russell, Stuart, 223; on 2018 "computer glitch", 199; on assault on human dignity, 198–99; beneficial AI approach of, 194–95, 206, 210, 215, 224–25, 228; on control problem of AI technology, 192, 193–94, 202–3, 206; on existential risk posed by superintelligence, 199–200; on the gorilla problem, 200–201; inverse reinforcement learning of, 196; on King Midas problem, 201; on need for cultural movement, 208; on symbiotic relationship between humans and machines, 210; warning on dependency on AI machines, 231n3

Ryan, Sharon: on theories of wisdom, 84–85; on wisdom, 76, 88

Sceptics/Scepticism, 28, 36, 63

self-control, 5–6, 30–31, 61, 62, 93–94, 127, 150, 159–60, 178–79, 180, 184, 203–4, 211, 221

self-fulfillment, 15; as capacity fulfilment, 37–38, 40–42, 55n17, 220; as neo-Stoic version of eudaimonia, 24n2, 29, 34–35, 40–42, 53, 55n22; transformative aspect of, 48, 66; and virtue, 45–46

Self-fulfillment (Gewirth), 29, 34, 37, 40, 41, 55nn8, 17

self-preservation, 48–49, 50, 51, 52, 66, 213, 227

self-respect. *See* dignity/self-respect

Seligman, Martin, 205

Sellars, John, 27

Seneca, 28, 30, 35, 208

Shakespeare, William: *Sonnet 116*, 230

shared responsibility, 165, 167–68, 172–73, 191, 222, 223–24

Sims, Rod, 171

Sisyphus (mythological character), 140, 141, 147n4

smart machines, 8, 191–92

sociability, 49, 50, 52, 66, 213, 227

social contract, 119, 171, 172; notion of, 123n12

social media. *See* 5th Estate

Socrates, 36, 43, 66, 84, 208, 221; on worthy life, 131, 132

Socrates in Love (D'Angour), 231n8

Socratic ignorance, 86

solidarity, 177, 182–83, 207, 212, 216; and equality, 182

"Some Moral and Technical Consequences of Automation" (Wiener), 194

Sorcerer's Apprentice (Goethe), 201

sovereignty, 154; digital, 150–52, 190; individual, 215–16; self, 160, 180

Space Odyssey 2001 (film), 195–96

Star Trek (TV series), 70, 71n4, 161, 179, 229

statistical correlation and causation, 153, 157–58, 161–62, 178–79, 182, 191

statistical dehumanization, 178, 179, 191

Stoicon conferences, 208, 231n4

stoic philosophy, 30; application to AI technology, 5, 210–14; application to technology, 2, 7, 59–60, 62–69; application to technology, relevance of, 70, 220; communal perspective of, 31–32; core principles of, 28, 30–34, 35–36, 63–69, 210–14; cosmic perspective of, 31, 36, 63–64; features in Gewirth's ethical theory, 29, 34–36, 53; indirect influences of, 27–28; influence on cognitive behavioural therapy, 205; influence

on professional ethics, 28; internal control in, 190–91, 203–4, 209–10; notion of love in, 227; problem of, 60–61; as therapeutic method, 60, 205–6; as a way of life, 32–34, 64–65, 68–69, 205, 208, 213–14, 220
Striker, Gisela, 27, 35; on nature, 52; notion of *oikeiosis*, 48, 49–50
Sundar Pichai, 151
superintelligence, 4, 156, 176; control problem posed by, 192–97, 199–200; motivation to comply and defer to man, 225–30
Superintelligence (Bostrom), 199, 206
supernature, 223–24
surveillance capitalism, 151; and privacy, 159–60, 181; we are the product in, 162, 180
surveillance practices, 62–63, 212; of media, 108, 115, 118–20, 125, 129; and Stoic cosmopolitanism, 64–65
sustainability/sustainable technologies, 13, 16, 20–21, 23
Symposium (Plato), 224, 225–27
systemic corruption, 116, 117, 118, 121, 125, 126, 151, 204

tech media corruption, 7, 101–3, 108–9, 114–15, 125; effects of, 115–19
technology: ethics of, 11–12; impact on well-being, 1–3, 6; impact on well-being, evaluation of, 7; and love, 224, 225–30; man, nature and, 219–21; man, well-being and, 221–24; problem of, 59–60; and stoic philosophy, 2, 5, 7, 59–60, 62–69, 70, 220
Tegmark, Max, 223; on control problem of AGI, 194, 206–7, 215; on goal-alignment problem, 196–97, 207; on importance of philosophy in AI development, 207–8
Tencent (company), 209
T-GLAT. *See* theory of the good life and technology

theoretical wisdom (*sophia*), 84–85
theory of the good life and technology (T-GLAT), 2, 11; adequacy of, 14–17, 22–23; meta-conditions for, 12–14, 15–17
Thompson, Dennis: on institutional political corruption, 118
Thunberg, Greta, 206
totalitarianism, 100
traceability of algorithms, 162–63, 165
transcendental freedom, 132
transparency, 3, 4, 5, 6, 7, 8, 62, 67, 119–20, 125, 149, 154, 155–56, 166, 168, 169, 181, 182–83, 207, 216. *See also* opacity
Trump, Donald, 109, 110, 117–18, 193, 202
trust, 3; in autonomous algorithms, 165–66; distinction between trustworthiness and, 156; fiduciary relationship of, 113, 114, 117; and media, 102, 103, 104, 107, 111
trustworthiness, 156
truth, 3, 7, 207; and media, 102, 103, 104, 107, 111, 119–20, 125; as necessary for information and knowledge, 77–78
Tufekci, Zeynep: on media corruption, 118
Turin, Alan, 223; on existential risk, 192, 193, 200
Turner, Tina, 224, 225
Twitter, 94, 122n1, 193

Umbrello, Steven and van de Poel, Ibo, 217
understanding, 83–84, 85, 87
United States presidential elections (2016): electoral interference in, 3, 110, 114, 117–18
universal morality, 38–39
universe: metaphysical perspective of, 138–39
utilitarianism, 207, 215

value alignment problem, 201
value-loading problem, 196–97
values, 80, 90, 97n8, 157, 167, 170, 171, 172, 177, 178, 181–82, 201, 215, 216–17
value sensitive design, 217
Van Dijck, José, Poell and De Waal: on challenges of enforcement agencies, 170; on shared responsibility, 167, 172
Véliz, Carissa: on breach of privacy, 67–68; on privacy and surveillance capitalism, 159, 181
virtue: and eudaimonia/happiness, 15, 28, 29, 82–83; and eudaimonia/happiness, Gewirth's views on, 45–46; and eudaimonia/happiness, Stoic's views on, 30–31, 32–33, 36–37, 42–45, 60–61; as final goal, 37; Gewirth's notion of, 45; and self-fulfillment, 37–38; Stoic and neo-Stoic method of control through, 203–5; Stoics' notion of, 42, 63, 211
Vlastos, Gregory, 24n4, 98n16
volition, 127, 137, 150–51, 159–60, 164, 191, 198, 203, 204, 209, 215, 221

Weckert, John, 11–12
well-being: AI technologies' impact on, 4–5, 7, 149–50; and autonomy, 183–85; ICT impact on, 7, 125–26; man, technology and, 221–24; and self-respect, 130–31, 132–34, 141–43, 144–46, 173–75; technology's impact on, 1–3, 6. *See also* eudaimonia
"What is Wisdom?" (Ryan), 76
WhatsApp, 105
"Why Rights are Indispensable" (Gewirth), 142
Wiener, Norbert, 194, 201
Williams, Bernard: on worthy life, 131
wisdom: eudaimonic notion of, 82; as evaluative principle, 18–20, 22, 23; as extensive factual knowledge (WFK), 84, 85; as knowing how to, and succeeding at, living well (KLS), 85, 88; as knowledge, 84–85; knowledge for, 85–87, 89; as link between information and well-being, 73–76, 80–83; as meta-technology of the mind, 64; as meta-technology of the self, 20; as meta-virtue, 73; notion of, 7, 18, 83–84, 88; psychology of, 88; relevance in normative evaluation of digital information, 90–96; Stoic's sense of, 30–31, 63–64, 211; theories of, 84–88; triadic notion of, 73, 74, 75; and well-being, 2–3

YouTube, 76, 93, 94, 102, 119, 122n4

Zeno of Citium, 35
ZiZek, Slavoj, 180; on Neuralink, 185
Zuboff, Shoshana, 101–2; on surveillance capitalism, 61, 120, 156
Zuckerberg, Mark, 107, 151, 173, 185; on regulation, 167, 168–70

About the Author

Dr Edward H. Spence, PhD (University of Sydney), is a philosopher specialising in moral philosophy and epistemology, as well as applied and practical ethics in the areas of computer ethics, ethics of technology and ethics of information. He holds research positions in the School of Communication and Creative Industries, Charles Sturt University, the Department of Philosophy at the University of Sydney, Australia, and the 4TU.Centre for Ethics and Technology in the Netherlands. Prior to that he was a senior research fellow of the Centre for Applied Philosophy and Public Ethics (CAPPE), Canberra, member of the Executive Board of the Association of Practical and Professional Ethics (APPE), USA, and member of the Executive Committee of the Australian Association for Professional and Applied Ethics (AAPAE). He is the author of numerous peer-reviewed papers in international journals, as well as the author of several books including *Media Corruption in the Age of Information*, (2021), *Ethics in a Digital Era,* with Deni Elliott (2018), *The Good Life in a Technological Age* (edited volume with Philip Brey and Adam Briggle (2012) *Media, Markets, and Morals, New Ethics for Digital Media,* with Andrew Alexandra, Aaron Queen and Anne Dunn (2011) *Ethics Within Reason: A Neo-Gewirthian Approach* (2006), *Advertising Ethics*, with Brett Van Heekeren (2005), *Corruption and Anti-Corruption: A Philosophical Approach*, with Seumas Miller and Peter Roberts (2005). He is the founder and producer of the *Philosophy Plays* project whose aim is the introduction of philosophy to the general public through drama performance and audience participation. Several of his philosophy plays have been performed at Arts and Cultural Festivals throughout Australia and the USA. He has contributed articles and talks on philosophy and ethics to various media publications, including the *Sydney Morning Herald, The Australian Financial Review*, and *Conversation, Neos Kosmos,* and *ABC Radio National*.

www.ingramcontent.com/pod-product-compliance
Lightning Source LLC
Chambersburg PA
CBHW021848300426
44115CB00005B/59